1/05

ALSO BY HENRY PETROSKI

Small Things Considered: Why There Is No Perfect Design

Paperboy: Confessions of a Future Engineer

The Book on the Bookshelf

Remaking the World: Adventures in Engineering

Invention by Design: How Engineers Get from Thought to Thing

Engineers of Dreams: Great Bridge Builders and the Spanning of America

Design Paradigms: Case Histories of Error and Judgment in Engineering

The Evolution of Useful Things

The Pencil: A History of Design and Circumstance

Beyond Engineering: Essays and Other Attempts to Figure Without Equations

To Engineer Is Human: The Role of Failure in Successful Design

Pushing the Limits

Pushing the Limits

New Adventures in Engineering

HENRY PETROSKI

ALFRED A. KNOPF   NEW YORK   2004

THIS IS A BORZOI BOOK
PUBLISHED BY ALFRED A. KNOPF

The essays in this book were originally published,
in somewhat different form, in *American Scientist.*

Library of Congress Cataloging-in-Publication Data

Petroski, Henry.
Pushing the limits : new adventures in engineering / Henry Petroski.
p.   cm.
Includes bibliographical references and index.
ISBN 1-4000-4051-5
1. Engineering—History. I. Title.
TA15.P47 2004
620—dc22   2003065845

Manufactured in the United States of America
First Edition

to Stephen

We were put on this earth to make things.

—W. H. Auden

Contents

Preface

Engineers are the masterminds behind most things made, from the miniaturized microelectronics device that fits into the ear to help the hard of hearing to the gigantic domed stadium into which a hundred thousand fans can crowd and cheer. Small things, large things, simple things, complex things—they are all the products of the minds of engineers and the hands of engineering.

Mainly, this book is a celebration of the bigger things that engineers make, the bridges and buildings and dams that are among the largest constructions on earth. The stories behind such achievements are often the stories of engineers with personalities as large as their plans, whose persistent dreams have become part of our reality.

Some of the greatest engineering achievements have been dwarfed by later ones. Thus, the first flight of the Wright *Flyer* was shorter than the wingspan of today's jumbo jets. And a pedestrian who in five minutes can walk between the towers of the Brooklyn Bridge—once the longest span in the world—would need twenty to walk from tower to tower of Japan's Akashi Kaikyo Bridge, today's longest. The title of greatest is one that seldom stands unchallenged for long, but even former champions still command a measure of respect. The Empire State Building, which reigned as the world's tallest for four decades, remains an awesome structure to which tourists flock.

Size is not the only criterion by which an accomplishment is judged. Great engineering projects, even those in the megaproject category, are not great simply because they are big. The ingenious creations and elegant solutions of engineers are also works of art and inspiration: an Eiffel Tower or a voyage to the moon lifts everyone's spirits. Above all, the stories of great structures and schemes are also the stories of people, often of those who persist in the face of opposition to realize what others had declared impossible.

Making things is an activity as old as civilization, and making ever new things is part of being human. Laypersons and engineers alike are forever encountering new problems, new needs, and new desires that challenge them to invent, to innovate, and to engineer still another thing, another scheme, another dream. The accomplishments of our predecessors provide both the stick and the carrot to keep us going—at the same time, we follow the tried-and-true ways and deviate from them. As footpaths evolved into roads, so roads developed into highways, but what the highways of the future will be like is unimaginable to most of us today.

The older is not always a reliable model for the newer, the smaller for the larger, or the simpler for the more complex, as countless failures and embarrassments throughout the history of civilization have demonstrated. Such temporary setbacks and disappointments are an integral part of progress, but they occur remarkably infrequently even in the face of the most difficult of challenges. Making something greater than any existing thing necessarily involves going beyond experience. This book is about such adventures in engineering: testing the boundaries, moving the frontier, pushing the limits.

Bridges

Art in Iron and Steel

Works of engineering and technology are sometimes viewed as the antitheses of art and humanity. Think of the connotations of assembly lines, robots, and computers. Any positive values there might be in such creations of the mind and human industry can be overwhelmed by the associated negative images of repetitive, stressful, and threatened jobs. Such images fuel the arguments of critics of technology even as they themselves may drive powerful cars and use the Internet to protest what they see as the artless and dehumanizing aspects of living in an industrialized and digitized society. At the same time, landmark megastructures such as the Brooklyn and Golden Gate bridges are almost universally hailed as majestic human achievements as well as great engineering monuments that have come to signify the spirits of their respective cities. The relationship between art and engineering has seldom been easy or consistent.

Arguably, the assembly-line process associated with Henry Ford made workers tools of the system, but Ford also wanted to produce automobiles that were affordable to working people, and he paid his own workers sufficiently well that they could save to buy the cars they made. The human worker may have appeared to be but a cog in the wheel of industry, yet photographers could reveal the beauty of line and composition in a worker doing something as common as using a wrench to turn a bolt. When Ford's enormous River Rouge plant opened in 1927 to produce the Model A, the painter-photographer

Charles Sheeler was chosen to photograph it. The world's largest car factory captured the imagination of Sheeler, who described it as the most thrilling subject he ever had to work with. The artist also composed oil paintings of the plant, giving them titles such as *American Landscape* and *Classic Landscape.*

Long before Sheeler, other artists, too, had seen the beauty and humanity in works of engineering and technology. This is perhaps no more evident than in Coalbrookdale, England, where iron, which was so important to the industrial revolution, was worked for centuries. Here, in the late eighteenth century, Abraham Darby III cast on the banks of the Severn River the large ribs that formed the world's first iron bridge, a dramatic departure from the classic stone and timber bridges that dotted the countryside and were captured in numerous serene landscape paintings. The metal structure, simply but appropriately called Iron Bridge, still spans the river and still beckons engineers, artists, and tourists to gaze upon and walk across it, as if on a pilgrimage to a revered place.

At Coalbrookdale, the reflection of the ironwork in the water completes the semicircular structure to form a wide-open eye into the future that is now the past. One artist's bucolic depiction shows pedestrians and horsemen on the bridge, as if on a woodland trail. On one shore, a pair of well-dressed onlookers interrupt their stroll along the riverbank, perhaps to admire the bridge. On the other side of the gently flowing river, a lone man leads two mules beneath an arch that lets the towpath pass through the bridge's abutment. A single boatman paddles across the river in a tiny tub. He is in no rush because there is no towline to carry from one side of the bridge to the other. This is how Michael Rooker saw Iron Bridge in his 1792 painting. A colored engraving of the scene hangs in the nearby Ironbridge Gorge Museum, along with countless other contemporary renderings of the bridge in its full glory and in its context, showing the iron structure not as a blight on the landscape but at the center of it. The surrounding area at the same time radiates out from the bridge and pales behind it.

In the nineteenth century, the railroads captured the imagination of artists, and the steam engine in the distance of a landscape became as much a part of it as the herd of cows in the foreground. The Impressionist Claude Monet painted man-made structures like railway stations (Gare Saint-Lazare) and cathedrals (Rouen) as well as water lilies. Portrait painters such as Christian Schussele found subjects in engineers and inventors—and their inventions—as well as in the American founding fathers. By the twentieth century, engineering, technology, and industry were very well established as subjects for artists.

American-born Joseph Pennell illustrated many European travel articles and books, including—among the many with his wife, Elizabeth Robins Pennell—*Over the Alps on a Bicycle*. Pennell, who early in his career made drawings of buildings under construction and shrouded in scaffolding, returned to America late in life and recorded industrial activities during World War I. He is perhaps best known among engineers for his depiction of the Panama Canal as it neared completion and his etchings of the partially completed Hell Gate and Delaware River bridges.

Pennell has often been quoted as saying, "Great engineering is great art," a sentiment that he expressed repeatedly. He wrote of his contemporaries, "I understand nothing of engineering, but I know that engineers are the greatest architects and the most pictorial builders since the Greeks." Where some observers saw only utility, Pennell saw also beauty, if not in form then at least in scale. He felt he was not only rendering a concrete subject but also conveying through his drawings the impression that it made on him. Pennell called the sensation that he felt before a great construction project "The Wonder of Work." He saw engineering as a process. That process is memorialized in every completed dam, skyscraper, bridge, or other great achievement of engineering.

If Pennell experienced the wonder of work in the aggregate, Lewis Hine focused on the individuals who engaged in the work. Hine was trained as a sociologist but became best known as a photographer who exposed the exploitation of children. His early work documented immigrants passing through Ellis Island, along with the conditions in the New York tenements where they lived and the sweatshops where they worked. His depictions of child labor in the Carolinas brought to public attention how young children toiled for long hours amid dangerous machinery. Hine depicted American Red Cross relief efforts during World War I and, afterward, the burdens war placed on children. Upon returning to New York, he was given the opportunity to record the construction of the Empire State Building, which resulted in the striking photographs that have become such familiar images of daring and insouciance. He put his own life at risk to capture workers suspended on cables hundreds of feet in the air and sitting on a high girder eating lunch. To engineers today, one of the most striking features of these photos, published in 1932 in *Men at Work,* is the absence of safety lines and hard hats. However, perhaps more than anything, the photos evoke Pennell's "wonder of work" and inspire admiration for the bravery and skill that bring a great engineering project to completion.

Alfred Stieglitz, who intended to study engineering at Berlin Polytechnic, redirected his interests to photochemistry after he acquired a small camera, and while still a student he began to work to gain recognition for photography as an art form on a par with painting. His early work showed steady technical innovation, as he took photographs in snow, in rain, and at night. He is considered the father of modern photography as an art form. In addition to making a series of four hundred prints of his wife, Georgia O'Keeffe, and also four hundred prints of cloud patterns related to emotions, Stieglitz captured with his camera memorable images of New York's Flatiron Building and other structures. (O'Keeffe herself, so well-known for her abstract floral forms and southwestern themes, painted views of the East River, dominated by rooftops and industrial smokestacks, and the Brooklyn Bridge that crosses that river.)

In the 1930s, Margaret Bourke-White, who established a reputation for photographing industrial sites, produced photo essays of Russia and the Soviet Union, and would go on to become the first official woman photojournalist to cover World War II. Some of her most widely known work was produced for *Life.* Her first assignment for the new publication was to photograph dams the Public Works Administration was constructing in the northwest United States, and she focused her efforts on the enormous (four-mile-long) Fort Peck Dam in northeastern Montana. Her famous portrait of the earthen dam's concrete spillway structure appeared on the cover of the inaugural issue of the magazine, dated November 23, 1936.

Edward Steichen, another pioneer in photography as an art form, was attracted to both the glamour of Hollywood (Greta Garbo and Charlie Chaplin were two of his subjects) and the squalor of the battlefield. He led the photography division of the Army Air Service in World

War I, and headed the U.S. Navy photography unit in World War II. As director of the photography department of the Museum of Modern Art, he organized the Family of Man exhibition in 1955. Steichen also photographed the Flatiron Building.

Joseph Stella, known for painting abstract themes (aquatic life and jungle foliage), returned throughout his life to the subject of the Brooklyn Bridge and often abstracted New York City in his paintings. The East River and Brooklyn Bridge also captured the imaginations of poets. In his "Crossing Brooklyn Ferry," Walt Whitman wrote about the river scene that so many commuters saw each day. He was one of them, and he reveled alike in the sunset and the ships in the harbor and the contrast of the foundry chimneys against the sky. When the Brooklyn Bridge superseded the ferry, it also succeeded it as an inspiration to poets such as Hart Crane, whose book-length poem *The Bridge* is perhaps the best known.

Although many painters, photographers, and poets have seen art and humanity in the products of engineering and technology, not all artists have. Many in the late-nineteenth-century Parisian artistic and literary community found the Eiffel Tower "an offense to good taste" and characterized it as coming from the "baroque, mercantile imaginings of a machine builder." The builder, Gustave Eiffel, defended his wrought-iron tower as "beautiful in its own right." He also defended the works of engineers generally: "Can one think that because we are engineers, beauty does not preoccupy us or that we do not try to build beautiful, as well as solid and long lasting structures?" He further held that "there is an attraction, a special charm in the colossal to which ordinary theories of art do not apply."

Indeed, an engineer designing a structure is not unlike an artist painting one. Both start with nothing but talent, experience, and inspiration. The fresh piece of paper on the drawing board is as blank as the newly stretched piece of canvas. And the greatest of bridge engineers, especially, have quite explicitly written and spoken of the aesthetic criteria and human values that influenced the shapes, forms, textures, and functions of their structures; the spans themselves stand as tributes to their successful applications of their ideals.

Iron Bridge was only the first in a long line of cast-iron, wrought-iron, and steel structures that have continued to grace the British landscape. Although its form was borrowed from the ancient lines of stone bridges and the details from the classic lines of timber construction, it did not owe its structural success to their principles. In particular, the strength of iron enabled subsequent bridges to be built with much lower profiles, thus marking them as daring and at the same time making them more user-friendly. Thomas Telford was one of the first masters of the shallow metal arch, with his 1814 Craigellachie Bridge over the River Spey near Elgin, Scotland, being an outstanding example. The crossed struts between the thin arch and the equally thin deck give the bridge a transparency and accessibility unknown in stone structures. The crenellated towers of the abutment, Telford's acknowledgment of the setting in which the bridge might otherwise have appeared to be an intruder, tie it into the culture of its place. Telford used a similar architectural motif on his Conwy Suspension Bridge, in deference to the Welsh castle to which it leads. Of course, Telford looked well beyond castles for his inspiration. His Menai Bridge married massive stone towers, which appear to have evolved naturally from the piers under the approach spans, with wrought-iron chains to produce a profile of near-perfect proportions that served as an aesthetic model for suspension bridges well into the twentieth century.

The Brooklyn Bridge, completed in 1883, is almost three times as large as the Menai, and during construction its towers dominated the New York skyline. The technical challenge that John A. Roebling faced in spanning the East River was to design a structure that would not interfere with shipping. This demanded not only a high roadway beneath which tall-masted ships could pass but also a great span to provide a wide, unobstructed channel. Suspension bridges were Roebling's forte, but the combination of constraints in New York called for one of unprecedented size. Rather than design a purely utilitarian structure, he produced a masterpiece. The tall stone towers pierced by the twin Gothic arches through which traffic passes are necessarily massive, but their monumental design makes them feel architecturally right. Roebling's patented steel-wire cables hang with a well-proportioned sag, counterpointed by the taut diagonal stay cables that the engineer included out of respect for the wind and what it could do to an unstayed bridge deck. The bridge deck was designed not only for horses and carriages but also for people, and the elevated walkway that puts the walkers above the road traffic makes the bridge at the same time a brilliant work of engineering, art, and humanity.

Brooklyn Bridge promenade

A walk across the Brooklyn Bridge is one of the world's great pedestrian experiences. The arched towers are triumphal, not in a military but in a civil-engineering sense. The cables and diagonal stays pull the eye upward to the top of the tower and to the prominent date stone embedded near the top. The high-altitude reminder that the Brooklyn tower dates from 1875 (the Manhattan one is from 1876) makes the walker feel not small and insignificant but part of a larger humanity that could erect this grand edifice with little more than muscle and steam power. To stroll the walkway of the Brooklyn Bridge is to experience the dynamic nature of the bridge itself, with the suspension cables first dipping down to meet the walker at midspan and then rising up again with the spirit to the tower top. The skyscrapers of lower Manhattan appear through the screen of steel as a great backdrop for the bridge itself. It is so grand in its execution that it is easy to forget that the bridge was built for the city, not the city for the bridge. Until recently, the twin towers of the World Trade Center dominated the view and appeared to be inspired by the bridge itself, echoing its twin arches in more modern lines. The bridge's empty arches now serve as reminders of what was once a different skyline, but seeing the sun setting behind a lessened skyline through the network of stone and steel is still as dramatic an event as watching it set over any mountain or sea.

The Brooklyn Bridge is an architectural masterpiece precisely because it is an engineering masterpiece. Its engineer was its architect. Although John Roebling assigned the preparation of its presentation rendering to an assistant engineer and better draftsman, Wilhelm Hildenbrand, the bridge's lines and proportions are all Roebling's. Yet in the half century after the completion of the Brooklyn Bridge, there was a regular tension between engineers and architects over who was responsible for designing bridges and what was needed to give their aesthetic details a monumental look.

Othmar H. Ammann, the engineer who was responsible for the greatest number of major bridges in New York City, retained professional architects to consult on aesthetics. Ammann's first independent bridge design was a suspension span to cross the Hudson River, and he engaged Cass Gilbert, the architect of the Woolworth Building, to render views of it to show to prospective supporters of the project. In the majority of Gilbert's studies of the bridge towers, they are clad in stone, which in the early twentieth century was still considered by many the appropriate way to express the great weight that they supported. In this case, the stone was to be only facade, however, applied over a steel skeleton that was the real means of support. Yet when the George Washington Bridge was completed in 1931, during the Depression, the steel had to be left exposed at first to save money. It was generally thought that the stone would be added when economic conditions improved. That did not happen, of course, and the architect Le Corbusier called the structure "the most beautiful bridge in the world," one on which "the steel architecture seems to laugh."

The single simple steel-arch motif—an echo of Iron Bridge—that Ammann incorporated high in the towers of the George Washington Bridge became his trademark on subsequent suspension-bridge towers that he designed. He did not wish to encumber his bridges with "extraneous architectural embellishments," but he did appreciate that even his preference for minimalist design did call for careful thought of shape and line. For the Bronx-Whitestone Bridge, he engaged the architect Aymar Embury II to work with Allston Dana, the engineer of design. The anchorages of the bridge especially have the look of being designed without being frilly. Their shape follows the lines of the suspension cables turning into the ground, and they give the bridge a streamlined look. In writing about his involvement in the design, Embury emphasized that although he and Dana had a free hand in the work, they were "always subject to Mr. Ammann's criticism and never out of his control." Embury left no doubt that the design was ultimately that of the chief engineer, and the bridge's structural function dictated its architectural form.

The Brooklyn and George Washington bridges may be among the most recognizable structures on the East Coast, but the Golden Gate Bridge is *the* most recognizable one on the West Coast. Although there has been considerable debate over how much credit chief engineer Joseph B. Strauss deserves for this masterpiece, there should be little doubt that he was the entrepreneurial driving force behind its construction. Strauss's first design for a bridge across the entrance to San Francisco Bay was a hybrid monstrosity with little but functionality to recommend it. It was the consulting engineer Leon S. Moisseiff who convinced Strauss to embrace a pure suspension-bridge form, one that would be the largest in the world when completed in 1937. The detailed design work was done by Charles Ellis, who was fired by Strauss when they disagreed over how much more calculation was needed for the structural design of the towers. Each of these engineers can be said to have contributed to the realization of the bridge in its basic proportions, which are defined by the height of the towers, the length of the span, and the sag of the main suspension cables.

Like Ammann, Strauss could ultimately have had the most to say about how the bridge looked. But Strauss seems not to have had the same aesthetic sensibility. His career had involved mostly the design and construction of drawbridges with no claims to grace. In fact, many of them were downright ugly. The extremely graceful Arlington Memorial Bridge in Washington, D.C., is sometimes attributed to him, but in fact he was only a consulting engineer on the one well-disguised openable span among its many fixed arches. Fortunately, when it came time to finish the Golden Gate Bridge, Strauss did not impose his own aesthetic sense, or lack thereof, on the structure. Rather, he hired consulting architects. The one that survived working with Strauss to the end was the local architect Irving Morrow. While the bare structure of the span does give it its essential lines, much of the appeal of the Golden Gate Bridge as a piece of art has to do with Morrow's architectural "details" of faceted fascia, sleek railings, and the color that the structure is painted. These finishing details add considerable interest and subtlety to what has been called the "world's largest Art Deco sculpture."

Today, one of the most innovative and influential engineers is Santiago Calatrava, who was trained also as an architect and who has been called an engineer-artist. His bridges and other structures show the influence of all three professions. At the same time, they provide public spaces of a human scale and stand as pieces of monumental sculpture in their own right. Thus, Calatrava, the subject of a later chapter in this book, has established himself as one of the most watched designers of large structures in the world. Increasingly, commissioners of bridges in America are looking to such individuals or to teams of engineers and architects, sometimes working also with artists, to develop concepts for the signature bridges that so many communities now desire. This growing awareness of the intangible added value of human space and art is sure to give us more masterpieces like the Brooklyn Bridge and the Golden Gate Bridge. They in turn will continue to be inspiring monuments to civilization and ever welcome subjects for artists of all kinds.

Bridges of America

America has more than half a million bridges, ranging from non-descript low-profile highway overpasses to monumental works that raise our cars and spirits to great heights, but only a few of them are known by name. Although just about everyone recognizes the Brooklyn Bridge and the Golden Gate Bridge and can tell you something about them, only local residents, attentive tourists, and bridge aficionados are likely to be aware of such structural jewels as Coos Bay Bridge, also known as Conde B. McCullough Memorial Bridge, and to know that Conde McCullough is the name of the structure's engineer and not that of a local Oregon politician. But whether or not bridges or their engineers are well-known, masterpieces of bridge engineering are everywhere, legacies of their designers' structural artwork that are as much a part of the American scene as are the red barns and white churches of the countryside or the tall buildings and canyonlike streets of large cities.

From Maine to California, from Washington to Florida, there are bridges of note, and to each belongs a story that is at once unique and typical. Many bridges are truly one-of-a-kind construction projects, and local constraints of topography, geology, traffic, politics, and economics make complications commonplace. Across America over the past century, local idiosyncrasies have led to the design and construction of bridges as odd and individual as many of their engineers. For example, the 1928 structure that carries Maine's Route 24 the thousand

13

or so feet between Bailey Island and neighboring Orrs Island is supported over most of its length on a crib of large and heavy granite ties that allows the tides to move freely in and out of the arm of Casco Bay that separates the islands. By designing a masonry structure for this location, the state bridge engineer, Llewellyn N. Edwards, obviated the problems of rot and corrosion that plague timber and steel bridges. Today, the Bailey Island Bridge stands as solid as ever, a monument to the foresight of its designer, who in his later years also looked backward to write a valuable history of early American bridges, thus preserving memories of so many classic structures and their engineers.

In a park near Kansas City, a modest span of one hundred feet peaks forty feet in the air, thus serving as a beacon for pedestrians looking to cross Rush Creek to a picnic area. The bridge was originally built in 1898 to carry a line of the Quincy, Omaha & Kansas City Railroad, until it was abandoned in 1930. It then stood unused for some years but after World War II was converted to local highway use. When the filling of the reservoir behind nearby Smithfield Dam threatened to submerge the bridge, it was disassembled in 1982 and its parts stored by the Army Corps of Engineers. The reconstruction of the bridge five years later was spearheaded by George Hauck, a professor of civil engineering at the Kansas City campus of the University of Missouri, who sought to preserve it as a monument to its engineer, J. A. L. Waddell, who patented the A-frame truss in 1894 and built scores of such bridges in the Midwest and Japan.

The Canadian-born John Alexander Low Waddell received his civil-engineering degree in 1875 from Rensselaer Polytechnic Institute, the premier engineering school in America at the time. After gaining experience with bridge-building companies and further study at McGill University, in 1882 Waddell accepted a faculty position at the Imperial University in Tokyo, which enabled him to revisit the Orient, to which he had traveled for his health as a sickly youth. After a few years in Japan, Waddell returned to America and soon started a consulting practice in Kansas City. The firm evolved over time to be known successively as Waddell & Hedrick, Waddell & Harrington, Waddell & Son, and Waddell & Hardesty, and lives on to this day designing bridges and more as Hardesty & Hanover.

From building modest A-frame trusses, Waddell's practice grew to the point where he was designing major bridges of many kinds. One of his firm's most unusual realized projects still spans the Missouri River in Kansas City. As originally designed, it carried railroad trains on a lower deck and highway traffic above. Because the bridge was built close to the water, it incorporated a lift mechanism to raise a section of the railroad tracks by retracting it into the upper structure so that the flow of highway traffic could continue uninterrupted. The lift mechanism for the bridge does not tower above the upper deck the way it generally does in vertical-lift bridges, and this Kansas City bridge is believed to be the only one of its kind extant. The structure is known today as the ASB—or Armour, Swift, Burlington—Bridge, after the meatpacking and railroad companies that once controlled it. Like many an old highway bridge, it had narrow lanes, and so was closed to vehicle traffic when the modern, higher, and wider Heart of America Bridge was completed nearby in 1986.

Portland, Oregon, is a veritable open-air museum of bridges designed by some of the most productive and influential engineers of the late nineteenth and early twentieth centuries. Local bridge historian Sharon Wood conducts fascinating tours of these structures and is a valuable repository of information about the city's bridges and their engineers. Two of Portland's bridges were built by Waddell & Harrington: the Hawthorne and the Steel. The former is a conventional vertical-lift bridge, with tall towers on either side of the movable span that serve as cranes to lift it out of the way of ships passing up and down the Willamette River. Completed in 1910, it is believed to be the oldest vertical-lift bridge still in full operation. The Steel Bridge, completed in 1912, has the telescoping feature of Kansas City's ASB Bridge, but it also has the conventional lift towers, allowing the entire movable span to be raised even higher out of the way. The name of the Portland structure dates from an earlier railroad bridge near the same location, which was built in 1888 when steel was displacing wrought iron as the railroad-bridge material of choice.

Other spans in downtown Portland are associated with some of the most innovative and prolific of American engineers, bridge designers who flourished during the period that reached from the heyday of the railroads through the ascendancy of the automobile. Among these engineers is Gustav Lindenthal, who began his career in Pittsburgh, where his Smithfield Street Bridge, completed in 1883, remains one of the outstanding examples employing a lenticular truss, whose lens-shaped profile makes the bridge easy to recognize. From Pittsburgh, Lindenthal moved to New York City, where for almost fifty years he promoted unsuccessfully his designs for a monumental bridge across the Hudson River. Lindenthal did, however, leave his mark on other New York spans, perhaps most notably the city's massive Hell Gate Bridge.

In Portland, Lindenthal is credited with three bridges: the Sellwood, the Ross Island, and the Burnside, all completed in the mid-1920s. This last is a double-leaf bascule (from the French for "seesaw"), a kind of drawbridge, with an opening mechanism designed by the firm of Joseph Strauss, who at the time was expending much of his energy in and around San Francisco promoting the design and construction of the Golden Gate Bridge. In Portland, another double-leaf bascule, the Broadway Bridge, was designed by Strauss's contemporary Ralph Modjeski, whose accomplishments ranged from modest-span movable bridges in Chicago to the world-class Delaware River Bridge (now the Benjamin Franklin Bridge) at Philadelphia.

Perhaps the most striking Portland structure is the St. Johns Bridge, a major suspension bridge distinguished by its Gothic towers and "verde-green" paint job, both choices of David B. Steinman, who began his bridge-building career working under Lindenthal on the Hell Gate. Unlike most large suspension bridges, the steel cables of which are spun in place wire by wire, those of the St. Johns were assembled out of fully formed lengths of wire-rope strands made at the John A. Roebling's Sons plant in Trenton, New Jersey. Steinman was responsible for bridges around the world, but like Lindenthal and many a prolific engineer, he did not realize all of his dreams. His proposal for a monumental "Liberty Bridge" across the entrance to New York Harbor was passed over in favor of what became the Verrazano-Narrows Bridge. That bridge was the design of Othmar H. Ammann, who had also worked under Lindenthal on the Hell Gate project, and whose 1931 George Washington Bridge across the Hudson realized the dream that eluded his mentor.

Though many an American city can claim, like Portland, to be a city of bridges, few areas of the country can rival the Oregon coast in sheer numbers of beautiful and unusual bridges. What distinguishes these structures as a group is the genius of their chief engineer, Conde B. McCullough, who was born in 1887 in South Dakota, studied engineering at Iowa State, worked for the Iowa State Highway Commission for a few years, moved to Oregon in 1916 to teach, and soon became the state's bridge engineer. In Oregon, he also studied law, wrote on economic and legal matters in addition to bridge design, and left a legacy of graceful bridges built in the 1930s to complete the Oregon Coast Highway. Although no two of McCullough's major bridges are exactly alike, virtually all contain his signature elements: graceful concrete arches and Art Deco entrance pylons.

Driving north from California on U.S. Highway 101, the first McCullough bridge encountered is also one of the grandest: the Rogue River Bridge at Gold Beach. The seven 230-foot reinforced concrete arches, the first in America built employing the method of prestressing developed by the French engineer Eugène Freyssinet, leap gracefully across an estuary. About seventy-five miles north, at the much wider Coos Bay, thirteen concrete arches flank a steel cantilever of major proportions and complement it nicely. This is the bridge named after McCullough, but all his bridges are memorials to him.

Seventy-five miles north of Coos Bay, approaching Waldport, one gets a spectacular view of Alsea Bay and the multiple-arch bridge cross-

Conde McCullough's Alsea Bay Bridge

ing it. In this location today, however, the traveler sees not an original McCullough bridge but an evocation of one. A modern replacement was erected in 1991, because the original structure was so damaged by the corrosive effects of saltwater on reinforced concrete. The new bridge, with its three arches above the road nicely providing a visual focus along the curving line of thirty-two 150-foot-span underdeck arches, was considered so significant a historic structure that the weakened and narrow bridge was allowed to be demolished and replaced only under special conditions. These included erecting a museum, known as the Historic Alsea Bay Bridge Interpretive Center, near the northbound bridge approach. Beside the southbound approach, a viewpoint for the new bridge is located in a pleasant park, into the landscaping and design of which are incorporated some of the original bridge's concrete pylons and part of its railings.

Many a bridge has a pleasant park in its shadow, often providing a fresh perspective on the scale of the structure. The Tacoma Narrows Bridge, across an arm of Puget Sound in Washington State, was rebuilt in 1950 with more attention to the aerodynamic forces that twisted the original span apart a decade earlier. One approach to the structure once straddled War Memorial Park, in which visitors could stroll among artifacts of battle and conflict and in the shadow of a massive anchorage. The true scale of a great suspension bridge can often be best appreciated when one stands beside such an anchorage, the size of which is dictated by the dead weight needed to counteract the tremendous pull of the cables. In late 2003, Tacoma's War Memorial Park became the construction site for another anchorage—for the new bridge that had to be constructed beside the existing one to accommodate increased traffic.

Other bridgeside parks provide more insight into the history of bridges than of battles and obsolescence. On the Kentucky side of the Ohio River near the suspension bridge to Cincinnati, a plaque describes the span and a statue depicts its builder, John A. Roebling, after whom the structure is now named. Near the toll plaza of the Golden Gate Bridge, there is a pleasant flower-filled park and a statue of Joseph B. Strauss. Beside it, a replica of a section of the structure's main cable shows how it is made up of individual wires, and this display attracts more attention than the statue of the engineer. Tourists seem pulled to the cable to touch it, as they are to a replica of the cables of the George Washington Bridge that stands near the Constitution Avenue entrance to the Smithsonian Institution's National Museum of American History in Washington.

Walking across bridges is another pleasurable way of appreciating their scale and the accomplishment they represent. A leisurely stroll across a bridge like the Golden Gate or the George Washington or the Brooklyn can take a good hour. Another hour for the return trip to where one's car is parked offers a reminder of the importance of a bridge's long approaches, which raise its roadway as high as a couple of hundred feet above the water. A walk across a great bridge can also provide a perspective on the size and height of its towers, which are akin to skyscrapers. Indeed, when the 746-foot-tall towers of the Golden Gate Bridge were under construction in the 1930s, they were the tallest structures in the San Francisco area. Today, they are the city's third tallest structures, just short of the Transamerica Pyramid and the Bank of America Building. In New York, the 690-foot-high steel towers of the Verrazano-Narrows Bridge, completed in 1964, remain taller than all but a couple dozen or so of that city's skyscrapers, and the curvature of the earth across the 4,260-foot distance between the towers makes them more than an inch farther apart at their tops than at their bottoms.

The Verrazano-Narrows and many other highway bridges built in the later part of the twentieth century do not allow for pedestrians. Such a practice feeds self-fulfilling prophesies that people today would rather drive than walk, thus justifying the elimination of "unnecessary" sidewalks on economic grounds. It is also sometimes said that the intent of walkless bridges is to discourage suicides, but nothing has kept cars from being abandoned by truly desperate drivers near midspan railings. The San Diego–Coronado Bridge in San Diego, which has no pedestrian access but does have suicide-hotline numbers posted on its approaches, arcs both vertically and horizontally in a long and graceful curve across the bay in order to achieve a considerable height over the main channel, thus allowing huge navy ships to pass beneath. This layout not only enables the bridge to fit the existing different street alignments of both the mainland and Coronado but also minimizes the need for approach viaducts on the peninsula itself, thus maintaining its residential character and low silhouette in southern California sunsets. It is a shame the sunset is not readily available for viewing from the bridge by pedestrians.

Every city has its distinctive bridges, whether they are oddly configured to satisfy local geometry and politics or curiously named to commemorate engineers whose accomplishments have been long forgotten by most users of the bridges. In any case, knowing the story of a bridge and its builder invariably reveals a rich and rewarding chapter in the history of a place, its people, and their dreams.

Benjamin Franklin Bridge

Cities and water go together. It was natural to establish early camps, settlements, and forts beside streams, rivers, lakes, and bays. Such locations provided ready access not only to the substance essential to life but also to fishing, transportation, and security. It is no accident that the nucleus of New York City was at the tip of lower Manhattan, which commands a strategic view of the harbor, or that the U.S. Military Academy sits on West Point, which is located at a double turn in the Hudson River, where enemy ships were forced to navigate slowly, thus making them easy targets.

As forts grew into towns and cities, waterways were both blessing and curse. Chicago's location on Lake Michigan gave it what at first must have seemed a near inexhaustible source of fresh water, and the Chicago River that flowed into the lake provided a natural sewer. However, the refuse eventually polluted the drinking water, producing a serious health hazard. The situation drove the city to undertake the enormous task of reversing the flow of the river, sending Chicago's sewage westward into the Des Plaines River and, eventually, into the Mississippi, which supplied water to other cities, including St. Louis. Fortunately, as predicted, the water was purified in the course of the long and turbulent journey.

Swift-running rivers and streams not only cleaned themselves but also provided power for mills. Yet the same flow that drove waterwheels complicated up-, down-, and cross-stream traffic, which was essential

21

for distributing the products of the mills. The continued development of a mill town into a city necessitated the construction of canals, locks, and especially bridges, the last of which was relatively easy for narrow crossings but a technological challenge for wider waterways. The growth of many a city was hampered as much by its lack of adequate bridges as by anything else.

The development of the railroad in the nineteenth century threatened the economic future of cities that could not provide bridges across wide rivers. Had the Eads Bridge not been built at St. Louis, virtually all long-haul rail traffic crossing the Mississippi River might have been routed through Chicago and northern Illinois, where the river was bridged, leaving the more southern city without significant commerce. The development of the automobile in the early twentieth century presented new challenges for bridge builders, requiring new spans to move traffic across previously unbridged waters.

Every river city has at least one bridge story, and Philadelphia is no exception. Philadelphia sits at the confluence of the Schuylkill and Delaware rivers, with the former being a tributary of the latter. The smaller Schuylkill was naturally the first to be bridged. The most famous of its early-nineteenth-century crossings was located at Fairmount, now a part of the city but then a bit outside Philadelphia proper. In 1813, Louis Wernwag completed a combination arch-and-truss covered bridge there, which had a clear span of 340 feet. At the time, it was the longest wooden bridge in America. According to David B. Steinman and Sara Ruth Watson in their classic *Bridges and Their Builders,* "it was a beautifully and originally designed bridge, worthy in all respects of its nickname," the Colossus. The great bridge has also been described by Donald C. Jackson in his indispensable guidebook, *Great American Bridges and Dams,* as "the most stunning and visually compelling engineering structure built in the early United States." Unfortunately, like many a contemporary wooden bridge, the Colossus was destroyed by fire—in its case, in 1838.

The Schuylkill was also spanned by two historically significant early suspension bridges, the engineers of which were the pioneers of the form in America. The first was the design of James Finley, who is generally acknowledged to have built the earliest iron-chain bridges. The second was the first of that type built by Charles Ellet, Jr. His bridge, with a 358-foot span suspended from wire cable passing over stone towers, was completed in 1842. It provided a replacement for the Colossus.

The much wider Delaware River was not bridged nearly so early nor so readily. An 1818 proposal for a bridge between Philadelphia and Camden, New Jersey, was unusual in that the span did not go all the way across the river. The scheme consisted of a bridge from New Jersey to Smith Island, which then existed close to the Pennsylvania shore, in combination with a ferry connecting the island and the foot of Market Street. The island itself was removed in 1893 to improve navigation, but other bridges reaching across the entire river were proposed even before that time.

There is always more than one way to cross a river, of course, and a multispan suspension bridge was proposed in 1851 by the Philadelphia-born engineer John C. Trautwine. Trautwine had gained extensive experience on railroad and canal projects, including surveying work on a railroad across the Isthmus of Panama, prefiguring the route of the future canal. His proposal for a bridge at Philadelphia with four spans of one thousand feet each would have been a daring stretch of the state of the art. Just two years earlier, Ellet's 1,010-foot suspension bridge had been completed over the Ohio River at Wheeling, West Virginia. Unfortunately, the roadway of that bridge was destroyed in a windstorm in 1854 and had to be reconstructed. Such a structural failure would naturally give pause to any community then contemplating a similar but larger bridge.

The collapse of the Wheeling span might have struck an even more severe blow to suspension-bridge building in America had not another bridge, designed by John Roebling, been completed at about the same time. His Niagara Gorge Suspension Bridge, with a clear span of 821 feet, was the first of that genre to carry railroad trains. Roebling's unique structure owed much of its success to his study of the effects of wind on large bridges. It was only by understanding how earlier suspension bridges had failed, Roebling believed, that he could design ones that would withstand the great forces of nature.

Roebling went on to design the great bridge across the Ohio River at Cincinnati. Though its construction was interrupted by the Civil War, the 1,057-foot span was completed and opened to traffic by 1867. The bridge at Cincinnati set the structural stage for Roebling's masterpiece, the Brooklyn Bridge, the construction of which was approved only because Roebling proposed a bridge high enough to give ample clearance to tall-masted river traffic.

Nevertheless, plans to cross the East River in New York with a span of almost 1,600 feet appear to have rejuvenated talk of a crossing of the Delaware River. An 1868 joint report by committees representing the interests of Philadelphia and of Camden put forth what they considered an improvement on the span planned for Brooklyn. The proposed structure was to be a low-level crossing throughout, incorporating a roadway that split into two drawbridge spans.

The idea was for the bridge to operate somewhat like a canal lock. When a ship wished to pass the bridge, traffic would be diverted to the far draw span, which would remain in the closed position. The nearer draw span would be raised, allowing the ship to pass into a wide pool between the two roadways. Then the raised span would be closed behind the ship, bridge traffic redirected once again, and the second draw span opened to give the ship free passage out of the pool. Sketches of the extravagant scheme were published, but not surprisingly the bizarre structure was never built.

But engineers did continue to dream up new schemes to bridge the great distance. J. A. L. Waddell was one of the most flamboyant personalities among bridge engineers of the time, and he produced an equally flamboyant design for crossing the Delaware. It called for a unique suspension bridge that answered some of the objections to bridges of this type. In order to avoid long land-based approaches requiring the acquisition of expensive land in established neighborhoods, Waddell employed compact helical approaches, not unlike the kind we encounter in some multilevel parking structures today. Such an unusual scheme won little support, however, and in 1919 Pennsylvania and New Jersey established the bistate Delaware River Bridge Joint Commission, under which careful and thoughtful planning would begin again in earnest.

The commission interviewed bridge engineers from around the country, and a three-member Board of Engineers was appointed. The board included George S. Webster, who had long been associated with public works in Philadelphia; Laurence A. Ball, who had extensive experience with railroad and subway work in New York City; and Ralph Modjeski, whose wide-ranging experience made him one of the preeminent bridge engineers of the time. It was Modjeski who chaired the board and who came to be named chief engineer of the project. The bridge was effectively to become Modjeski's bridge.

Modjeski, who was born in Poland in 1861, first came to America in 1876 with his mother, Helen Modjeska, an actress who was described as the "premiere tragedienne of her time." The family visited New York and Philadelphia, where they took in the Centennial Exposition. Young Ralph no doubt noticed that there was no bridge across the Delaware, as he did that there was no canal across the Isthmus of Panama, which they traversed by rail. His doting mother later recounted that the teenage Modjeski declared that "someday he would build the Panama Canal."

After studying engineering in Paris at the prestigious Ecole des Ponts et Chaussées, Modjeski returned to America in 1885 to begin his engineering career. He first worked under George S. Morison, who has been called the "father of bridge building in America," but soon opened his own office in Chicago. Modjeski's name would eventually be associated with more than fifty major bridges, including bascule, truss, arch, cantilever, and suspension types. If he would not build the Panama Canal, he would build bridges that were also world-class achievements.

Modjeski became a member of the board of engineers charged with overseeing the redesign of a massive bridge across the St. Lawrence River at Quebec, the first version of which collapsed during construction in 1907. The Quebec Bridge is an enormous cantilever structure, which means that its spans were built like outstretched arms, supported only at one end. It was finally completed in 1917, and, at 1,800 feet between piers, it then had the longest single span of any bridge of any kind anywhere in the world. But the failure of the first Quebec Bridge so shook the bridge-building community that the cantilever form soon fell out of fashion for the longest spans, and the Quebec Bridge remains to this day the longest of its kind ever built.

No doubt Modjeski's association with such an ultimately successful record-setting achievement made him an attractive candidate to lead the effort to build a bridge across the Delaware River. Though the cantilever form was still considered an option, at least with Modjeski in charge, in the end a suspension bridge was determined to be the less expensive alternative. According to *Engineering News-Record*, the leading chronicler of the construction industry, in a 1921 article reviewing engineering studies for a Philadelphia-Camden bridge:

> The advantage of superior rigidity possessed by the cantilever bridge was believed to be negligible in highway service. The cantilever saves a large item of cost in simpler anchorages and approach construction. The suspension bridge, on the other hand, is not only very much lighter, but was considered to involve less risk in construction, to be adapted to subdivision of contracts, to require less time to build, and to be cheaper in maintenance on account of having less metal exposed. It was also considered to have the esthetic advantage over the cantilever bridge.

Indeed, when completed, at 1,750 feet it would be the longest suspension span in the world. It should not be surprising that an engineer with the background, talents, and ambitions of Modjeski would enthusiastically embrace the opportunity to build another record-breaking bridge.

Regardless of experience or talent, no single engineer or board of engineers, no matter how distinguished, can alone build a world-class bridge. There is too much specialized knowledge required, there are too many design alternatives to be considered, and in the final analysis there are simply too many details to be worked out. Even with the decision to build a suspension bridge across the Delaware, there remained many subsidiary technical choices to be made before construction could begin. These included exactly where the bridge should cross the river, where the towers should be founded and how high they should be, where the anchorages should be placed, what the line of the cables should be and whether they should be of eyebars or wire, what the configuration and proportions of the roadway should be, where train tracks and walkways should be located, what the exact dimensions of each piece of stone and steel should be, and what kinds of architectural details should be included. Consulting engineers and architects are engaged to help address such issues. Among the principal consultants for the Delaware River Bridge were Leon S. Moisseiff and Paul P. Cret.

Moisseiff served as consulting designing engineer to the board and later as engineer of design for the project. He was, in fact, to have a hand in the design of virtually every major suspension bridge built in the three decades after the 1909 completion of the Manhattan Bridge, to which he had applied the so-called deflection theory. This theory, by taking into account the interaction of the cables and roadway of a suspension bridge, enabled design engineers to make a more accurate determination of how forces were distributed among the various parts of the structure. With this knowledge, the steel components could be designed more optimally and hence the bridge built more economically.

Before such calculations could be made, however, the overall proportions of the structure had to be established. There are no rigid formulas for doing this; it involves a combination of working within the constraints of the bridge location and local geology and of making aesthetic judgments about what just looked right. In taking this all-important design step of setting the proportions of a major structure, engineers act more like artists than scientists. And only after the overall defining geometry is set down can the theories and formulas of engineering science be applied to the details. Modjeski and Moisseiff were each masters of both the art and the science, but as chief engineer Modjeski would have the final word and responsibility for the way the bridge looked.

Although a suspension bridge is the purest of structural forms, needing no decorative treatment or facade to give it an aesthetic presence, architects are often involved to recommend finishing touches. Modjeski engaged Paul Cret as architect to provide advice on details large and small. Cret had considerable input on such aspects of the bridge as the exterior design of the enormous anchorages that it required, and the stone pylons that frame the steel towers for everyone driving across the bridge. Yet for all of the thought that went into the materials, proportions, and appearance of the structure, the artist Joseph Pennell would still call the bridge under construction over the Delaware "the ugliest bridge in the world." That judgment was reinforced by the aging master bridge engineer Gustav Lindenthal, who considered the towers to be designed "too much on the utilitarian principle of braced telegraph poles or derricks, holding up ropes." Such opinions were disputed by many other observers, who found the structure beautiful.

Whatever the aesthetic verdict on the bridge, from a purely engineering point of view it was an innovative and record-breaking achievement. Modjeski's board submitted its defining recommendation for the location and type of structure on June 9, 1921, and it was accepted by the joint commission just two weeks later. The total cost of the bridge was estimated to be almost $29 million, with the State of New Jersey responsible for about $12.5 million and the Commonwealth of Pennsylvania and the City of Philadelphia responsible for just over $8 million each. The board declared that the bridge could be finished for the country's sesquicentennial celebrations being planned for July 4, 1926. It would open three days early, almost fifty years to the day from when Modjeski first came to America. The event took place as planned despite construction delays caused by a debate over whether tolls would be charged. In the end, tolls were used to pay for the project, which was believed to be the largest public-toll enterprise up to that time.

The span, width, and traffic capacity of the bridge required cables of unprecedented size. Each cable was spun in place and was made up of twenty thousand individual steel wires, which meant the spinning operation had to handle almost twice as many strands as on any previously built suspension-bridge project. When compacted, the finished cables measured thirty inches in diameter, which made them half again as large as any then built. The use of new and stronger materials in the cables and lighter materials in the roadway reduced the overall weight of the structure, as did the use of lightweight trusses. These things in turn kept the overall cost down.

For all of the attention to technical innovations, the bridge was also designed with full regard for the people who would use it. Vehicle lanes were located in the center of the bridge, flanked and guarded on either side by sets of train rails and trusses. One set of tracks ran inside and one outside each of the bridge's two stiffening trusses. Over the outside tracks, cantilevered out from the top of the trusses, was a pair of pedestrian walkways. As on the Brooklyn Bridge, the elevation of the walkway above the road and rail traffic improved safety and gave pedestrians spectacular views uninterrupted by passing cars or trains.

Benjamin Franklin née Delaware River Bridge

Engineering News-Record celebrated the completion of the Delaware River Bridge in an editorial entitled "An Engineering Monument." The bridge was said to be "worthy of admiration most of all as an embodiment of modern engineering skill." The editorial went on to note that the Delaware River Bridge, coming nearly half a century after the Brooklyn Bridge, in a sense represented the concluding chapter in the book of suspension-bridge building. Whatever open questions in engineering knowledge remained after the completion of the Brooklyn Bridge, the Delaware River Bridge was believed to have closed.

Its reign as the longest suspended span in the world lasted only three years—until 1929, when the main span of the Ambassador Bridge between Detroit and Windsor, Ontario, reached across a distance fifty feet greater—but the Delaware River Bridge was by far the more attractive bridge and remains so. The 1931 completion of the George Washington Bridge, with a main span exactly two times that of the Delaware River Bridge, ushered in a new era—one of longer, sleeker, and lighter structures. Unfortunately, within a decade engineers like Moisseiff had allowed their hubris to get the better of their judgment, and the nadir of suspension-bridge building was reached in the 1940 collapse of the

Tacoma Narrows Bridge. Fortunately, the Delaware River Bridge had none of the inherent flaws of the Tacoma Narrows or related bridges. Now known as the Benjamin Franklin Bridge (and more familiarly as the Ben Franklin), the structure stands as a lasting monument to Ralph Modjeski and to the role of civil engineering in forging vital links across the waters between neighboring cities.

Floating Bridges

Whatever its type, a bridge is designed to carry something over some obstacle—a road, a valley, a river, a lake. Bridging a lake can be among the most challenging problems an engineer might face.

Walden Pond is not large enough to be called a lake, but plumbing its depths beneath the ice one winter gave Henry David Thoreau plenty of insight into the nature of things large and small. His 1846 drawing of the pond, which he reproduced in *Walden,* showed it to be no more than about 150 rods long by 100 rods wide. He also measured the depth of the pond along its major and minor axes and found it to be deepest—about one hundred feet—near where they crossed, a fact from which he generalized about the character of men. (Thoreau tended to use the surveying units of rods—each of which equaled 16.5 feet—for horizontal distances and the more familiar feet for vertical distances.)

At the time Thoreau surveyed Walden Pond, the state of the art of bridge building in America and the world would have allowed a single-span suspension bridge to be thrown across the narrowest part of the pond—where it was only fifty rods wide—but not across its widest part. Such a distance would not be bridged in a single span until the Brooklyn Bridge was completed in 1883. Bridging the length of Walden Pond with two spans would also have pushed the limits of mid-nineteenth-century engineering, for setting a foundation would have required working at depths greater than those at which workers digging the foundations of the Eads and Brooklyn bridges would experience the then-mysterious "caisson disease" now understood to be the bends.

If even a body of water as small and placid as Walden Pond could not easily have been bridged in the mid-nineteenth century, then how did the Persian king Xerxes throw a bridge across the Hellespont—the strait between the Gallipoli Peninsula in Europe and Turkey in Asia—in order to invade Greece almost twenty-five centuries ago?

The solution then, as it could have been at Walden Pond, was to use a floating bridge. In its simplest form, such a bridge is simply a boat. In many a crowded harbor or even at a busy dock, where boats tessellate the water by being jammed from stem to stern and from port to starboard, a common way to reach an outer one is to walk across the inner ones. This was how, a few years ago, a party of engineers I was with boarded a riverboat at Sandouping to begin a journey up the Yangtze River and how we disembarked at several busy cities on our way to Chongqing.

When a boat or ship is free of its moorings, it is metaphorically a bridge, carrying its passengers from one point of land to another. Ships are structures that float, and as such they must be designed to withstand the forces to which they will be subjected. The first test of a vessel's strength traditionally came at launch, when it slid down the ways and for the first time felt the force of buoyancy. With the stern supported in the water and the stem still on the ways, the hull was literally a bridge between water and land, and many larger ships were not up to the task of carrying even their own weight. The breakup of timber ships upon being launched was long known to be a danger but still not understood when Galileo mentioned the phenomenon in his 1638 treatise on the then-new engineering sciences of strength of materials and dynamics.

Even if they survived launching, ships were not out of danger. Floating vessels on a rough sea are bridges spanning the crests of waves, and as such are as much subject to bending as is a slender plank thrown across a construction ditch. Indeed, it is in such a situation that the hull of a ship can be subjected to the greatest stresses since launch and if not properly designed can break up.

Such extreme conditions are not often met on inland lakes or smooth-flowing rivers, and the use of anchored boat hulls with planks spanning between them can serve as a floating bridge, as they might have for Xerxes. If the boats are large enough and are connected not by planks but by substantial girders and decks, a bridge of some capacity can be assembled rather easily and quickly. Such was the kind of bridge proposed by Gustav Lindenthal to cross the Hudson River at New York as an interim measure while his design for a suspension bridge of enormous proportions was under consideration.

Pontoon bridge across the Hudson River, as proposed by Gustav Lindenthal

Floating bridges were used in the early nineteenth century on Lake Champlain, between upper New York State and Vermont. Known as draw boats, they were able to be moved out of the way to allow cross-traffic to pass. I have heard that these draw boats survive only at the bottom of the lake, where they can still be seen under appropriate con-

ditions of light and water clarity. A modest floating bridge still functions on a dirt and gravel road near Brookfield, Vermont. Though partially covered with water when my wife and I crossed it one spring, the
bridge remains a vital link in the local road network, at least during

Vermont floating bridge

nonwinter months. A sign near the structure across Sunset Lake
informed us that the first floating bridge at this location was built by
Luther Adams and his neighbors in 1820. The present bridge, the seventh to span the long and narrow lake here, was built in 1978 by the
local Vermont transportation district. This suggests that the average life
of the previous floating structures was about twenty-five years, which is
near the age of the present bridge, which presumably floated higher on
the water when new.

Pontoon bridges have long been used by the military, in the tradition
of Xerxes. The U.S. Army built many such multispan bridges during

World War II, some as long as 1,200 feet. During the war in Bosnia in the mid-1990s, muddy conditions along a rising Sava River delayed the completion of a pontoon bridge that was to carry peacekeeping troops into Bosnia. What bridges and roads had survived the fighting would have failed under the weight of seventy-ton tanks and heavy guns being moved into place, and so the construction of a floating bridge was essential to the mission. When the weather permitted, twenty-four-foot-long, six-ton folded sections of steel and aluminum were dropped into the river by a helicopter and pushed into place by boats. After eighty-five pontoon sections were in place, the largest floating bridge erected since World War II was ready to carry troops and equipment across. A proof-test of sorts was conducted when an eighty-eight-ton tank-recovery vehicle—one capable of towing disabled tanks—made the crossing. The bridge groaned and sank a bit under the weight, but it performed its function—until it was disassembled that night.

Challenging construction projects are not limited to wartime, of course, and often they take place under more tolerable conditions to produce structures of a more permanent kind. A large but shallow body of water can be crossed relatively easily with a solid bridge set on firm foundations. Thus, the seventeen-mile-long Chesapeake Bay Bridge-Tunnel consists mainly of a series of modest bridge spans set on piles driven into the bottom, with the bridge traffic carried into tunnels at two strategic points to maintain unobstructed shipping lanes. Twenty-four-mile-wide Lake Pontchartrain, near New Orleans, is spanned by a similar low-level structure, without the tunnel sections, as is the Seven Mile Bridge that carries traffic to Key West, Florida.

When a wide body of water is extremely deep, however, it is not always practical to construct the number of deep foundations needed to support even a long-span bridge, and so if a fixed link is desired alternatives have to be sought. The alternative of choice is often a permanent floating bridge of such substantial proportions that to the uninitiated driving across it seems no different than driving on a more conventional type of bridge. The mass and measure of such a floating bridge can offer substantial resistance even during storms, and so the choice of structure is one of economics and function.

The first modern civilian permanent floating bridge of any substantial size is believed to have been completed in 1912 across the Golden

Horn at Istanbul. The Galata Bridge employed fifty steel pontoons to span about 1,400 feet of water, the depth of which reaches 125 feet. Shortly after a new, parallel bridge was opened beside it, the Galata Bridge burned and sank. It was, however, rebuilt and is in service today.

Major permanent floating bridges are often essentially large ship hulls joined stem to stern and paved with what functions as a continuous roadway. Since such a bridge presents a solid barrier to cross-water traffic, shipping lanes have to be designed into the structure, which is usually done by spanning gaps between floating sections with a truss or other standard bridge type or by providing a movable span of some kind. The Galata contains a swing bridge section. (The first floating swing bridge powered by electricity was apparently constructed in the 1890s at Northwich, England, on the River Weaver Navigation.)

Conditions conducive to floating bridges exist in Lake Washington, which forms the eastern boundary of Seattle and separates it from the city of Bellevue. The lake, which is connected to Puget Sound by a ship canal, is almost twenty miles long and up to a couple of miles wide, with no significant currents or ice floes. As is suggested by the hilly topography of Seattle, the glacially carved lake is also deep. Its average depth of about 150 feet presented considerable challenges to any engineer proposing to bridge the lake.

Homer M. Hadley was born in Cincinnati in 1885 but moved to the West Coast as a surveyor with the U.S. Coast and Geodetic Survey. When time and location permitted, he studied engineering at the University of Washington but did not receive a degree. During World War I, he worked in Philadelphia building concrete ships and barges for the emergency fleet, the unorthodox material being used because there was a shortage of steel. Hadley returned to Seattle after the war, and in 1920 he suggested the use of concrete pontoons to support a floating bridge across Lake Washington. His proposal became public when he presented the idea at a meeting of the American Society of Civil Engineers, and considerable debate ensued.

The idea of a floating bridge on the scale proposed by Hadley was criticized for its lack of aesthetics by the navy, which had a station at Sand Point, on the lake. Bankers, calling the scheme "Hadley's Folly," ridiculed his proposal to use private money to be paid back with toll revenue. But such an idea for financing large bridge projects was in the air and would make possible the near-contemporary projects of the George Washington and Golden Gate bridges. Nevertheless, in the face of the opposition Hadley's idea became dormant. He took a job with the Portland Cement Association, in which he was charged with promoting the use of concrete in large-scale construction projects. He went on to design an early road-paving machine and become a prominent member of the trade association, all the time keeping the floating-bridge idea in mind.

Among the people to whom Hadley tried to sell his scheme was Lacey V. Murrow, director of the Washington State Department of Highways. The coincidence of Hadley's idea being on the table when federal highway funds became available prompted Murrow to instruct his staff to study the feasibility of a floating bridge. The idea was found sound, and Murrow wished to go ahead with it, but he wanted Hadley to assume a low profile regarding the project. Since the Portland Cement Association's motto was "to extend and promote the uses of concrete," Murrow feared that any prominent involvement by Hadley would give opponents of the bridge too easy a target.

Although still considered a radical approach, the construction of what would be the world's largest floating bridge—to Mercer Island, in the southeast corner of Lake Washington—was approved in 1937 and first crossed by traffic in 1940. The 1.4-mile-long bridge, consisting of a couple dozen pontoon sections about three hundred feet long, was an instant and enormous success and opened up development of the east side of the lake. At first known as the Mercer Island Floating Bridge, in 1967 (coincidentally the year of Hadley's death), it was renamed the Lacey V. Murrow Floating Bridge. Murrow received this extraordinary recognition in part because he had not kept his promise to Hadley that his role in promoting the bridge would not be forgotten and that in time he would be given credit. However, in spite of Murrow's reneging, Hadley's efforts were in the end recognized, for when a second, parallel bridge to Mercer Island was opened in 1989, it was named the Homer M. Hadley Floating Bridge.

Lake Washington is also crossed by another 1.4-mile-long floating structure, Evergreen Point Floating Bridge, which opened in 1963 and crosses the lake farther north. Still another floating bridge crosses the Hood Canal, which had been an obstacle to traffic headed to the

Olympic Peninsula, located west of Seattle. The 1.5-mile-long Hood Canal Floating Bridge, which opened in 1961, shortened driving access to some points on the peninsula by as much as one hundred miles.

It was actually the failure of the nearby Tacoma Narrows Bridge two decades earlier that had in part prompted the choice of a floating bridge across the Hood Canal. But floating bridges are not without their own problems and challenges. Among the latter are joining a series of pontoons end to end and keeping them in place by anchoring them individually to the water's bottom with cables that may be more than a mile long. In Seattle, joining the pontoons into a continuous longitudinal structure was chosen over the military preference for transverse pontoons because the former, while presenting more resistance to waves, resulted in less strain on the superstructure carrying the roadway.

Another great challenge to floating-bridge designers is how to accommodate tides and shipping lanes. The Hood Canal bridge, for example, crosses water that has sixteen-foot tides, which would make it difficult to maintain a smooth connection with the land if the bridge rose and fell with the tides. The situation was dealt with by mounting the roadway on piers attached to pontoons submerged below the lowest tide. The piers are high enough that the roadway is above the highest tide. A shipping channel is provided by means of an unusual drawbridge arrangement. Instead of a movable span that lifts or turns to allow water traffic to pass, the Hood Canal bridge has a pontoon section that moves longitudinally from its closed position into a split pontoon, not unlike a boat moving into a slip.

River currents can also present challenges to designers of floating bridges, as they did in Tasmania, Australia, where the Derwent River enters Hobart Harbor. After visiting Seattle, where the Mercer Island bridge had just been built, representatives of the Tasmanian Highway Department decided to use a floating bridge to carry traffic across the Derwent, anchoring one end of it to a lift bridge to allow shipping to pass. To deal with the strong river current, they designed a floating bridge in the form of a hinged horizontal arch, so that the current tended to stabilize rather than displace the bridge. The near-parabolic plan of the bridge was almost its undoing. Shortly after the completion of the bridge, a storm generated upriver waves that were reflected off the bridge and became concentrated at the focal point of the arch geometry. This concentration of waves at one point reflected ever stronger waves back toward the bridge, almost destroying it. After that experience, a barrier of pilings was constructed at the focal point to break up any future waves that might become focused there.

Floating bridges, put in place precisely because of the demand for convenient connections, normally carry heavy volumes of traffic. When the second Mercer Island floating bridge was being planned, residents of the island wished that the traffic was out of sight and the noise out of earshot. Anticipating Boston's Big Dig, Interstate 90 was covered over on Mercer Island with what is locally referred to as "the lid." Thus, instead of an open busy freeway creating an eyesore and earaches, it carries cars in tunnels beneath land given over to quiet park space and attractive recreational facilities.

No matter how the traffic is dealt with on the land, it is problems on the water that dominate design considerations for floating bridges. Among the most important design criteria is, of course, that water be kept out of the pontoon sections of a floating bridge. This has proved to be among the greatest problems experienced with the Seattle structures. In 1990, a year after the second Mercer Island floating bridge opened, the original one, closed for reconstruction, sank in a storm. The failure was attributed to workers having left hatches open over the Thanksgiving holiday, when the storm occurred, so that the pontoons of the unattended bridge filled with water and began to sink. Cracks in the pontoons aggravated the situation, allowing more water to enter. The failure was said to have proceeded in a falling-domino fashion. When it was rebuilt, the bridge incorporated prestressed-concrete pontoons, which put them in compression and thus closed any cracks that might initiate. In addition, more watertight cells were incorporated into the new pontoons, to confine any flooding that might develop. Seattle had seen an earlier failure when the western half of the Hood Canal Floating Bridge sank in a storm in 1979. The failed portion of the strategic bridge, which did not have a parallel companion span to which traffic could be diverted, was rebuilt within four years. The Evergreen bridge has also been battered in storms, resulting in cracks developing in the pontoons. But all bridge types are subject to damage and require maintenance, and Seattle's floating bridges remain the best solution to traffic needs in the context of the area's distinctive topography.

The Seattle-area floating bridges remain among the relatively few permanent pontoon structures in the world. Two notable ones were built in the 1990s in Norway. The Bergsoysund Bridge, which is a curved structure that sits on seven prestressed-concrete pontoons, spans more than 2,600 feet. The 3,800-foot-long Nordhordland Bridge crosses the Salhus fjord, which is more than 1,500 feet deep. The majority of this structure floats on transverse pontoons, and the shipping channel is spanned by a cable-stayed bridge, a bridge type that was promoted in the mid-1950s by Homer Hadley, who called it a "tied-cantilever," and that was just being introduced in Europe. But Hadley was again ahead of his time, for it was to be another two decades before a cable-stayed bridge would be built in America. Perhaps fittingly, the first such bridge in the contiguous United States was erected (in 1978) over the Columbia River, between Pasco and Kennewick, in Washington, where Hadley left his mark as an engineer. Locally, the Pasco-Kennewick Bridge is known simply as the "cable bridge."

A significant floating bridge was put into service in 2000 in Osaka, Japan, and it solves the problem of maintaining a wide shipping channel in an unusual way, even for a floating structure. The floating portion of the bridge, which is more than 1,200 feet long (with a clear span of more than 850 feet), looks not unlike a conventional steel span supported near its extremes on two massive piers. The bridge is in fact supported on two hollow-steel pontoons. When a ship wants to pass, which is not expected to be often, the entire bridge is rotated to the side of the channel. A smaller and much older floating bridge operating on a similar principle is located in Willemstad, Curaçao, in the Netherlands Antilles. Known as the "Floating Lady" and the "swinging old lady," this bridge dates from 1888 and still opens about six times a day. Since 1996, another modest floating bridge has provided a convenient means for pedestrians to walk between Canary Wharf and the West India Quay in London's Docklands, on the Isle of Dogs. Though spanning less than three hundred feet, this elegantly designed structure is a model for how attractive a floating bridge can be.

Currently, the technology of floating bridges is being combined on the drawing board with that of offshore oil and gas exploration, in which it is not uncommon to tether massive drilling rigs in water depths of more than one thousand feet. Among bridge proposals that rest on floating foundations are those designed to cross the fifteen-mile-wide Strait of Georgia, near Vancouver, British Columbia; the sixteen-mile-wide Strait of Gibraltar; and between pairs of Hawaiian islands, where water depths can exceed two thousand feet. Such ambitious crossings are likely to be among those discussed in the coming decades, but like Homer Hadley's first Seattle floating bridge, they are not likely to be realized until the right combination of circumstances—technical, economic, and political—arises.

Confederation Bridge

F ew large engineering design and construction projects are undertaken without opposition or controversy. Their completion will
alter on a grand scale the way things are, and whether such
changes are for better or worse often depends on one's point of view.
Thus, in the early part of the twentieth century, San Francisco's plan to
build a dam across Hetch Hetchy Valley—in the Yosemite National Park
region of the Sierra Nevada—was opposed by the environmentalist
John Muir and the Sierra Club, which took its opposition all the way to
the U.S. Supreme Court. Although the dam was built and today consti-
tutes a vital component in the infrastructure that provides water to the
city on the bay more than 160 miles away, opposition remains. There
continue to be calls to demolish O'Shaughnessy Dam, as it was named
after the city engineer who was the driving force behind its construction, and restore the valley to its pristine beauty. Not all battles between
technological developers and environmental conservationists reach
such epic proportions, but countless examples of large bridge projects
do come close.

In the context of water crossings, what are known as fixed links are
permanent structures, such as bridges or tunnels, that are much less
susceptible to the vagaries of the weather than are ferryboats. Thus,
fixed links provide more reliable means of communication across a river
or between an island and the mainland. Where there are narrow and
shallow waterways in heavily populated and trafficked areas, fixed links

have generally developed with the economy and society, for they were relatively easy to construct with contemporaneous technological experience and economic conditions. Often, however, where cities and other significant population centers have long been developing on different sides of wide and deep waterways, dreams of fixed links fall within technological and financial reach only after considerable human-centered development has taken place. Then they frequently conflict with the strong social and cultural milieu that has evolved, not to mention the established infrastructure of business and ferry interests, and the streets and buildings that would have to be closed, demolished, or displaced to make way for a fixed crossing.

The saga that culminated in the Channel Tunnel between England and France spanned almost two centuries. Opposition to that fixed link had much more to do with political and environmental issues and xenophobia than with technological difficulty. British fears of invasion by armies of soldiers and rabid animals from the Continent did more to keep the tunnel from being built than anything else. In America, a bridge across the Mississippi River at St. Louis was little more than a dream of civic boosters before the strong personality of James Buchanan Eads prevailed and went forth with the design and construction of a fixed link able to resist not only the forces of the river itself but also those of ferryboat and steamboat operators, rival Chicago business interests, and a vindictive leadership in the U.S. Army Corps of Engineers, which oversaw the waterways. Had Eads not prevailed, St. Louis might have lost its commercial viability and ceased to grow.

For more than a century, opposition and controversy as fierce as any in recent history surrounded the dream of a world-class bridge over the ice-forming waters of the Northumberland Strait between the mainland of Canada and Prince Edward Island. The country's smallest province, this 250-kilometer-long island is nestled beside the southernmost shores of the Gulf of St. Lawrence just off the coasts of two other maritime provinces, New Brunswick and Nova Scotia. The area is known for its crops of fine lobsters, scallops, and potatoes. It is also known for being the setting of the *Anne of Green Gables* novels of the early-twentieth-century local writer Lucy Maud Montgomery, a fact that has been turned into a significant island industry. That *Anne of Green Gables,* Canada's longest-running musical, was sponsored by the eventual bridge-development company and advertised on its user-friendly Web site is an indication of how important good public relations are believed to be for a fixed-link project.

When P.E.I., as the island is informally known, agreed to join the Dominion of Canada in 1873, it secured from the federal government several concessions. Although a summer-only steam ferry had run between Charlottetown, P.E.I., and Pictou, Nova Scotia, as early as 1832, in exchange for becoming part of the confederation islanders gained the assurance of "efficient steam communication between the island and the mainland." At first, this took the form of "primitive, open iceboats" that were not effective in heavy ice or storms, but in 1877 the government in Ottawa agreed to provide a subsidy for more comfortable steam service. Still more reliable year-round service came in the winter of 1918 with the introduction of ice-breaking ferries. Most recently, diesel-powered ferryboats provided year-round regular transport across the narrowest part of the strait very near where the bridge was finally built, but even in the late twentieth century ferry schedules remained subject to interruption by such conditions as ice and high winds.

The movable links with the mainland that ferries provide to an island are frequently subjects of debate. Those who treasure an island culture, relishing its remoteness and isolation and even taking a kind of pleasure in the long, slow ferry ride home from more "spoiled" locations, see a boat ride as a desirable transition between the hectic and the bucolic. In contrast, there are those who are in a hurry to get back and forth between home and business, who have a truckload of perishable goods to take across the water, who grow frustrated with the wait for ferries that may sail only after bad weather has passed or moderated, and who curse the winter and the ice it brings as well as the summer and the tourists and seasonally long ferry lines it brings. Opponents of fixed links tend not to be overly concerned with economic growth, whereas those in favor are likely to see them as panaceas for ailing economies. On P.E.I., the former wished to see their island stay much as they knew it; the latter saw a fixed link as an invitation to greater tourist traffic and a greater participation in the kind of economy that the mainland enjoyed.

Fixed-link proponents generally have an advantage in that they can pursue technical and financial possibilities with little or no formal authority. They need put forth a proposal only when they have a feasible one that they believe has a chance of winning whatever government or popular support is necessary for its implementation. It is often only then that opponents to a dry crossing can react, put forth their objections, and rally against the construction of a link. Depending on the inclinations of those in political office at the time, proponents or opponents then will have an edge.

Although a fixed link in the form of a tunnel under the Northumberland Strait was proposed as early as 1885, the most serious initiatives surfaced only in the second half of the twentieth century, when the P.E.I. economy was ailing badly. The island population, which stood at about 110,000, was experiencing unemployment rates considerably above the Canadian national average, and other economic indicators were equally worrisome. To address the problems, the government proposed a series of reforms, which included a fuller exploitation of agricultural resources and the development of tourism as an industry. The use of ferries to get produce and tourists from and to Prince Edward Island was believed to be an impediment to economic growth.

In the mid-1960s, the decade old idea of building a causeway over the relatively shallow strait again came under serious consideration. A causeway can be formed essentially by dumping earth and rock in the water to form a high and dry narrow strip of land on which a road can be constructed. However, such a fixed-link option across the Northumberland Strait could not maintain support as experience accumulated on the effects of the massive Canso Causeway in Nova Scotia on nearby benthic fisheries—those occurring at the bottom of the bay, inhabited by shellfish. Between 1955 and 1975, lobster catches in Chedabucto Bay declined by 95 percent, and it has been estimated that the resultant loss to the provincial economy was as much as one hundred million dollars (Canadian). Because of such controversy, any kind of fixed link was rejected by the federal government in favor of an economic-development agreement and improved ferry service.

Then in the mid-1980s, with a Conservative government in Ottawa, there "appeared from nowhere" three "unsolicited private-sector proposals," including both bridge and tunnel schemes, for a fixed link between P.E.I. and New Brunswick. Not unexpectedly, considerable public debate resulted, but when islanders voted in a plebiscite in January 1988, 60 percent of them endorsed going ahead with a fixed link. Opponents of a bridge later claimed that those voting for the link understood that a tunnel was being considered as a viable option. In the meantime, consultants had been commissioned to study the economic, structural, and financial viability of a fixed link; construction companies had been qualified to submit proposals; and public consultation had been initiated to consider the environmental and socioeconomic implications of likely schemes. By mid-1988, seven formal proposals were received by the Department of Public Works, including one for a tunnel. To the disappointment and anger of bridge opponents, only bridge proposals were finally accepted as meeting the necessary criteria.

The following year, a federal environmental-assessment panel was appointed to study the concept of a bridge across the Northumberland Strait. After extensive public hearings, the panel rejected a bridge as a fixed-link solution and suggested that a tunnel or improved ferry service be considered instead. Within three months, the Department of Public Works overruled the environmental review, however, arguing that its objections to a bridge could be overcome. A private re-review was then commissioned, and the minister of public works finally determined that a bridge should pose no unacceptable environmental risks and called again for the finalist proposals to be evaluated against environmental requirements and financial criteria.

Among the issues of greatest concern to environmentalists and fishermen alike was that of ice-out, as the breaking up and floating away of winter ice in the strait is known. Opponents of a bridge feared that delaying ice-out for more than a couple of days even once in a hundred years, something it was feared that a bridge might cause, could alter in a significant way lobster, scallop, and herring catches in the region. Early in 1992, however, the minister of public works reaffirmed that the bridge proposals did meet environmental requirements, and the successful proposal was to contain the bid that incorporated a need for the lowest annual federal subsidy. A low bid that amounted to a subsidy of forty-two million in 1992 Canadian dollars gave the project to Strait Crossing Development, a consortium of three Canadian companies. The government subsidy was a key component in the private financing of the bridge.

The project was to be completed under what is termed a design, build, finance, operate, and transfer arrangement. Strait Crossing was to assume all costs and financial responsibility, provided that it was allowed not only to have the federal subsidy that would normally go to the ferry service but also to keep all tolls collected on the bridge for a period of thirty-five years, at which time the bridge would be transferred to the government as the owner, maintainer, and operator. The government saw this as a good deal because it did not have to find the $840 million needed to build the bridge, no doubt at the expense of other projects, and did not have to continue to subsidize a ferry operation, including the anticipated acquisition of new capital equipment. Strait Crossing was protected against inflation and rising operating costs, but only to a degree. Opponents saw the government as being taken and as eventually inheriting a thirty-five-year-old bridge, whose hundred-year design life would present uncertain maintenance and, ultimately, decommissioning costs.

Although the story of the Confederation Bridge, as it was eventually named, is clearly one laden with, if not dominated by, politics and economics, it was also one of the most ambitious engineering projects ever undertaken. At eleven thousand meters, plus another 1,800 or so meters of approach spans, it is one of the longest water-spanning bridges in the world.

To address the environmental issue of ice-out and to provide sufficient clearance for shipping, as well as to keep construction costs down, the bridge was designed with as few piers in the deeper water as technologically feasible and with a typical vertical clearance of forty meters, rising as high as sixty meters above the water at the navigation channel. The massive concrete piers of the main bridge were fabricated in two main parts designed to be fitted together at their resting place in the strait. The pier bases rest in dredged recesses directly on the bedrock at the bottom of the strait. A pier shaft, which is the visible part of the structure, was slipped over the top of the pier base, and the entire assembled bridge was designed so that its sheer weight keeps it in place. As protection against the forces of ice, the pier shafts have incorporated into their design ice shields in the form of conical skirts, up the exterior slope of which the ice rises and thus breaks under its own weight. The bridge superstructure is of a cantilever design, which means that balanced atop the piers are 190-meter-long precast reinforced hollow-box-section concrete spans, which are also posttensioned—that is, fitted inside with taut cables that not only compress the concrete to obviate cracking but also assist in carrying the load. A sixty-meter gap between the tapered cantilevers was filled by dropping in a sixty-meter suspended span to produce a continuous and visually graceful profile across the wide strait.

The construction of a bridge with such massive components took special engineering considerations. To avoid having to erect cofferdams or employ caissons in the hostile and annually frozen waters of the strait, all pier and span components were prefabricated in a construction yard on P.E.I. Special crawler vehicles capable of lifting and transporting completed concrete components as heavy as 8,200 metric tons carried them to a landing, where they were picked up by the heavy-lifting vessel *Svanen,* which took them out into the strait and set them precisely in their ultimate locations with the aid of a global-positioning system. (The HLV *Svanen* was used originally to put heavy girders in place on the Great Belt Bridge connecting Denmark and Sweden. After that project was over, the *Svanen* had its height increased to one hundred meters and its twin ninety-four-meter-long pontoons widened for greater stability and buoyancy. Then the vessel was towed across the Atlantic on a submersible barge to do service in the Northumberland Strait.)

Construction on the new fixed link began in October 1993, with the preparation of the sixty-hectare construction yard and staging facility. Beginning in the summer of 1995, ferry passengers between Cape Tormentine, New Brunswick, and Borden, Prince Edward Island, watched the towering *Svanen* transport bridge-pier and girder components between the construction yard and their assigned places in the strait. The bridge, which has high and solid concrete parapets to keep acrophobic drivers from freezing at the wheel midway across, follows a gently curving route so as to address another psychological problem, that of drivers becoming hypnotized while traveling down a monotonous thirteen-kilometer straight and narrow concrete channel.

Throughout the construction period, on the ferry ride to P.E.I., itself a sinuous route from dock to dock, regular and first-time passengers alike tended to list the vessel by congregating portside so they could view the bridge in progress, with its initially isolated piers first joined by gapped cantilevers and eventually by completed spans. Early in the project, some ferry riders gazed wistfully at the still wide and untouched stretch of water in the middle of the strait. Only as the ferry curved into the small harbor at Borden, with the construction yard off to starboard, did the scale of the bridge become evident. Pier bases and shafts, cantilever and drop-in girders, in varying stages of production, were aligned in rigid rows with the concrete tracks and rail lines along which they would be carried by the turtlelike crawling machinery to be loaded on the *Svanen*. The construction of the bridge was the talk of the ferry and the island, and guided tours of the construction yard were popular with residents and tourists alike.

Naysayers remained fearful of what would happen to the area fishing industry. The strait ice broke up nicely during the winter of 1996, the first year any piers were in place, but the installed piers were then still few, and the winter was unusually mild. There remained some appre-

Confederation Bridge

hension about how the bridge would affect the ice and the environment generally in the long term. This remains to be seen, say those who still would have preferred a tunnel or no fixed link at all to P.E.I., but early indications were that there was little to worry about.

The bridge was completed and operational in late spring 1997, at which time Strait Crossing began collecting its subsidy and tolls. Tourism was hoped to increase by as much as 25 percent in response to the reality of the fixed link, and the island economy was expected to be revitalized. Opponents continued to be pessimistic, pointing to the hundreds of jobs lost by the closing of ferry service, but on P.E.I. there were new jobs collecting the round-trip toll, which in 2003 stood at $38.50 for an automobile. Not a few tourists began to go to the island simply to see the bridge and experience crossing the strait in ten to twelve minutes. It remains to be seen, however, if this will continue as the bridge becomes taken for granted as just another part of the infrastructure. What does not remain in doubt is that for the foreseeable future Confederation Bridge will remain the longest bridge over ice-covered waters in the world. It will attract pontists—as many bridge buffs tend to call themselves—on "bridging" trips, as the photographer Robert Cortright calls his excursions to admire and photograph the structures that make that travel itself easier and more pleasurable.

Pont de Normandie

O n a sunny spring Saturday in Honfleur, the small French sea-coast town on the left bank of the Seine estuary, it was no easier to find an empty table in a restaurant than it was to find a parking space on the narrow streets. It is not that there were too few restaurants or too few parking spaces in this town of ten thousand, for Honfleur's economy has long relied heavily on tourists and on accommodating their appetites for fine French seafood. The crowds were larger than usual this day, however, because of the newly completed bridge, the tall but slender towers of which could be seen, incongruous yet unobtrusive, beyond the rooftops across the harbor. After a leisurely meal of, most likely, fish and sauce and wine and cheese, not a few of the parties of visitors drove the few kilometers along the road through fields where cows graze lazily to the parking lot beside a construction office that had finished serving its purpose. Some parked their cars to walk halfway across the bridge and back, while others drew lots to see who would drive across to wait for the pedestrians on the other side. Either way, the three- or four-kilometer walk took a good hour to complete, but it was a beautiful day and a beautiful bridge—and for the time being at least it was the longest span of its kind in the world.

The Pont de Normandie, like virtually every large bridge, is at once a distinct solution to a unique engineering problem and, in a larger historical context, Everybridge. In 1959, when the Pont de Tancarville was opened, about twenty kilometers inland, it provided the only fixed crossing of the Seine between the coast and Rouen. The Tancarville bridge was at that time among the ten largest suspension bridges in the world, with a main span of 608 meters (about the size of the contemporary Walt Whitman Bridge in Philadelphia, the central span of which is two thousand feet long), and it remains today a beautiful and impressive structure with austere towers of gray concrete, hexagonal cables of exposed black steel strands, and a deep deck truss painted bright red, as if to taunt the wind. The Tancarville crossing was a very welcome improvement in road connections from the south to the major seaport of Le Havre, on the right bank of the Seine estuary, but it was not very long before the movers and shakers of that city promoted an even closer river crossing, although this would require an even larger bridge. There appeared to be no technological impediment to that, however; the quarter-century-old Golden Gate, with more than twice the span of the Tancarville bridge, provided all the precedent one could want. But before any bridge can be built there are financial as well as technological prerequisites that must be satisfied, and the financial climate of the 1970s forced plans for a new crossing to be laid aside.

With the concern that the absence of a crossing at the mouth of the Seine would leave Le Havre and its left-bank counterparts such as Honfleur high and dry on the discontinuous coastal tourist and commercial route, around 1980 the Chambre de Commerce et d'Industrie du Havre relaunched its campaign for a bridge. After conceding that tolls would have to be part of the financial planning, bridge boosters found sufficient regional support to go ahead with the project. Local departments of transportation are not expected to have on hand the engineering expertise to design world-record bridge structures, however, and the project was able to proceed only because the central French government provided design help through the Service d'Etudes Techniques des Routes et Autoroutes (SETRA), the technical arm of the French roads directorate. Serious design work for the new bridge began in the fall of 1986, under the direction of Michel Virlogeux.

The designs of all world-class bridges have begun as the dreams of individual engineers whose confidence in their own aesthetic and structural judgment matched the confidence of their clients in their financial and political judgment. Such engineers are a special breed. Michel Virlogeux was born in Vichy in 1946 and attended the world's oldest school of civil engineering, the prestigious Ecole des Ponts et Chaussées in Paris, where he later taught. His first engineering job, beginning in 1970, consisted of a three-year stint of national service on loan to the roads and bridges administration in Tunisia, across the Mediterranean Sea from France. In the meantime, he continued his studies and earned a doctorate from the University of Paris, his thesis being on the stability of columns that yielded under loads. In 1974, he joined SETRA, just as the organization was being given increased control over bridge design because many bridges that had been built under contractor-implemented alternative designs were found to be cracking. By 1980, when Virlogeux became head of large bridges for the organization, SETRA had established itself in a new position of power and influence in bridge design. His guiding principle, that "quality came from simple clean designs," informed his work on fifteen cable-stayed bridge designs at SETRA, but only two of these were realized before the Pont de Normandie.

Chambers of commerce may know how to promote local dreams and generate moral and financial support for a large bridge project, but they are seldom equipped with the technical expertise to judge the practicality of a design or to oversee its construction. This is the role of a project manager, and for the bridge across the Seine this role was filled by Bertrand Deroubaix, who headed a district transportation office when he was tapped for the big project and whom the French press had recently christened M. Normandie for his key role in revitalizing the Normandie project in the mid-1980s. It was Deroubaix who asked Virlogeux, who as a younger engineer had worked on the tower design for an earlier five-hundred-meter span at the site, to develop a new bridge design. Extensive studies were carried out to reduce costs and come up with a viable scheme for a crossing. With this done, in 1988 a law was passed in the French Parliament formally enabling the Le Havre chamber of commerce to be the owner of the bridge and to serve as a client to direct the development of a design and then to seek a construction contract.

When looking to have a bridge constructed, the standard French practice had been to present to contractors what is called an *avant-projet détaillé*. This preliminary detailed project design was typically based on an overall concept, dimensions, and structural calculations, but contractors were free to propose alternatives with their bids. When it did not result in cracked structures, this practice all too often resulted in unimaginative designs that involved applying construction procedures with which the contractors were already familiar to situations that might warrant, if not demand, unique solutions. Because of the clearly unusual nature of the Pont de Normandie project, its *avant-projet* was much more detailed than usual, especially with regard to analysis for dynamic conditions. The engineer Virlogeux, furthermore, wished to retain much more control of his design than was expected. It was possible for him to do so because of the confidence Deroubaix had in him and because of his own strong personality.

As is not unusual with novel engineering projects, an independent review panel was established to assess the safety and practicality of the structure and to recommend changes in the design. However, in his 1993 report on the design and construction of the Pont de Normandie, Virlogeux made clear his control over the project by writing that the panel's recommendations were "considered" rather than adopted. The fact that the chief engineer's extensive report was published while the bridge was still under construction was noted by other engineers to be unusually welcome, for such reports typically follow by years the completion of a project, by which time there is naturally less interest in them. The report's publication itself was thus a further indication of Virlogeux's high degree of confidence in himself and in what could fairly be called his bridge.

Although Virlogeux did not diminish the contributions of the members of his design team or the benefits gained from the reports of the bridge projects of other engineers, it was clearly his personality that dominated the Normandie operation. Colleagues, coworkers, and associates over the years remarked, some more euphemistically than others, that in their dealings with Virlogeux they found him to be "very direct" and, in fact, sometimes "to speak his mind too readily." He was recalled by engineers at SETRA to be an "exciting, stimulating person, who is nevertheless difficult and dogmatic at times," and to be one who "doesn't listen very well." According to the German bridge designer Jörg Schlaich, a member of the review panel, Virlogeux was "happy to have what he thinks confirmed, but not to have people come up with alternative proposals." Although this clearly made him difficult to deal with,

at the same time "this is his strength." Schlaich further understood that
with bridge building generally, "If you listen to too many people, you
will never come to the point. This self-confidence is what you need" to
design a world-record bridge. In this regard, Virlogeux is in the tradition of great bridge builders such as John Roebling.

The site of the Pont de Normandie is a major shipping channel, and
introducing piers into the water not only would have posed obstacles to
oceangoing vessels but also would have promoted the accumulation of
sand, a further obstacle to shipping. This led Virlogeux to look to piers
in excess of eight hundred meters apart, which at the time put the projected central span well beyond cable-stayed experience and into the
range of suspension bridges, such as the one upstream at Tancarville.
Thus, a suspension bridge would have been the expected form to adopt
at Le Havre. However, Virlogeux found that, although the pylons or
towers of a cable-stayed design would be more expensive because of
their additional height, stay cables could be installed at less cost than
suspension cables, and there would be a significant savings in not having to construct massive anchorages. On the other hand, the cable-
stayed solution did entail construction in a realm that was unknown. To
counter this uncertainty, Virlogeux did what engineers before him have
long done—that is, he pointed out that in the past great bridges had
successfully been built with spans that went well beyond experience. In
1931, for example, Othmar Ammann's George Washington Bridge was
completed with a 3,500-foot suspended span, which nearly doubled the
previous record 1,850-foot span of Detroit's Ambassador Bridge.

Because Virlogeux knew that he was designing in a new realm, he
looked carefully at suspension-bridge technology for guidance. In particular, he studied the deck designs of two British suspension bridges,
the Severn, with a span of almost one thousand meters, and the Humber, with a span in excess of 1,400 meters, then the longest in the world.
These streamlined hollow box-girder decks were developed in the
course of conducting wind-tunnel studies after the Tacoma Narrows
failure by the British firm of Freeman Fox & Partners. The satisfactory
behavior of those bridges, along with the cost savings associated with
the lighter box-girder structure, led to the deck design of the Pont de
Normandie, with its record main span of 856 meters—or 2,808 feet,
almost exactly that of the ill-fated Tacoma Narrows Bridge.

The new bridge's box-girder deck is further innovative in that it is of a hybrid construction. The approach spans and the 116 meters of the central span cantilevered out from each pylon are made of concrete, whereas the remaining 624 meters of the central span are of steel. This design was developed for various reasons, including economics; different construction demands of different parts of the bridge; the need to balance construction and permanent loads on the structure; and the

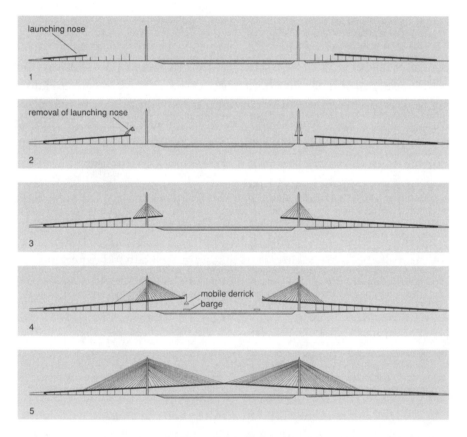

Construction sequence for the Pont de Normandie

need to take into account the stiffness of the structure in the wind, especially during the latter stages of construction when the free ends of the main span would be most susceptible to movement. Such unusual con-

ditions led to some disagreements over who among the various contractors was responsible for guaranteeing the overall safety of the structure and also led, subsequently, to concerns for the soundness of the
project. These concerns were, of course, largely dismissed with the successful completion of the bridge.

Accurately addressing questions relating to the aerodynamic stability of the bridge, both during construction and after completion, was
central to the success of the project. During construction, it was necessary to steady the incomplete bridge not only to keep it from being
overstressed but also to enable lifting and welding operations to proceed under controlled conditions. The use of steadying cables anchored
in the water would have interfered with shipping, so a tuned mass
damper system, which checked unwanted motions, was employed in
the free ends of the bridge deck to steady it until the final section of the
deck could be inserted, an event that took place in mid-1994.

The aerodynamics of the stay cables proved to be of even greater
concern. It was known that long cables could be susceptible to vibrations induced by anything from people walking on a bridge to the wind
blowing across it. In the latter case, the effects are aggravated in the
presence of rain running down the long cables. The cables of the Pont
de Normandie are enclosed in cylindrical ducts of high-density poly-
ethylene, the exterior surface of which is scored with helical grooves to
break up rivulets of rainwater that might form aerodynamically undesirable patterns. Excessive amplitudes of vibrations of bridge cables can
lead to premature fatigue damage in the wire strands, so measures had
to be taken to check any vibrations.

Even in the absence of rain and in modest winds, however, the
longer cables of the new bridge exhibited significant vibrations, so
alpine climbers were engaged to install damping ropes, which effectively tie the cables together. Since the cables have different lengths, they
have different natural frequencies; thus, coupled cables work against
each other to damp out vibrations. The six largest cables in each central
arrangement on the Pont de Normandie have also been retrofitted with
damping devices that connect them directly to the bridge deck. These
two dozen devices were hastily designed and constructed out of large
welded-steel links and heavy-duty truck shock absorbers, and they have
to be considered a blemish on an otherwise sleek and elegant structure.

Although the misbehavior of the cables in the completed bridge was unwelcome, it cannot be said to have been fully unexpected. Each structure of the scale of the Pont de Normandie is unique, and as such its behavior remains a hypothesis on paper until the structure itself is built. There is, of course, considerable testing of the hypothesis in the form of computer and scale modeling and calculation, but the results of all such testing depend heavily on modeling assumptions and calculational accuracy. Before the electronic computer, completed structures were customarily proof-tested with real loads to observe load-deflection behavior, which was checked against design calculations. But such real-world tests have increasingly been called unnecessary, if not downright damaging, by computer modelers, who argue that their numerous calculations are test and proof enough.

As is still customary in France, however, the Pont de Normandie was subjected to a traditional proof-test by having four lanes occupied by eighty fully loaded trucks parked nose to tail over 320 meters of the main span so engineers could observe the bridge-deck deflection under about 70 percent of its maximum working load. The extraordinariness of the bridge was underscored by the specification of two more unusual dynamic proof-tests. In the first, the midspan of the bridge was tied to a ship anchored below, and the cable was tightened to one hundred tons, at which point it was designed to break. The effect was the same as dropping an equivalent weight from the middle of the bridge to set it into free vibration. In the second dynamic test, a tugboat pulled the bridge deck sideways with a force of eighty metric tons and released it suddenly to set up transverse vibrations. The resulting amplitudes of vibrations were checked against calculations.

Such engineering details were surely not on the minds of the pedestrians who outnumbered the vehicles on the bridge on the sunny spring Saturday afternoon when I visited it. Few of the people seemed even to notice the windsock flying almost horizontally on the approach to the bridge proper or the sign warning parents to hold the hands of their children. Indeed, on the leeward walkway, there was hardly any wind to be felt, and the two or three meters separating the walkway railing from the edge of the bridge, which itself has another, lighter railing, made people forget that they were sixty meters above the Seine, which stretches out with a great expanse into the English Channel to the west.

To the strollers, the cables seemed as steady as the massive concrete pylons, and there was no perceptible movement of the bridge by wind or traffic. Indeed, there was so little traffic that it was possible to walk out to the center guardrail and capture the full symmetry of the bridge in photographs. Of the few automobiles that were crossing the bridge, more than one stopped to let out a passenger who wished to walk the remaining distance or to allow its driver to get out and snap a photo. The bridge was as welcome a promenade as the Jardin des Tuileries.

Whether enough cars would pay the toll of thirty-two francs (almost seven dollars) to help cover the cost of the newly opened billion-franc bridge remained to be seen, but traffic was bound to pick up with the tourist season and as word of the bridge spread. When that happens, those arriving in Honfleur too late in the day will no doubt find it next to impossible to find a parking place or a table for lunch. The people on the walkways of the bridge are likely to spill over onto the roadway and compete with automobiles for space in the slow lane. If the Pont de Normandie will be found by those pedestrians to have a flaw, it will not be in the vibrations of the cables or the damping devices that have been installed to suppress them; rather, it will be in the walkways that are too narrow to accommodate the crowds that the bridge is bound to attract. The annual Open Day for the structure drew thirty thousand visitors in 1993, before it was even possible to walk across it.

Whatever the future pedestrian and vehicle traffic on the Pont de Normandie, however, it is likely that few laypersons would recognize Michel Virlogeux if they saw him standing at midspan touching a cable to check on its movement. His name will long be remembered among French engineers, however, for as his successor at SETRA has said, "If he had not been there, the bridge would not be there." Where he himself would be in the future was to depend on the commissions he could attract as an independent consulting engineer, for he left the French civil service on January 21, 1995, the day after the Pont de Normandie was formally opened, to seek still greater freedom in designing other new bridges.

Britannia Bridge

A group portrait painted by the artist John Lucas shows an assembly of engineers and contractors gathered around the nineteenth-century English engineer Robert Stephenson, whose hand rests on a set of plans spread before him. In the background is an imposing structure, a partially completed bridge constructed of huge iron tubes. It is unlike any that any other engineer would design then or now. Yet this bridge across the Menai Strait between northwest Wales and the Isle of Anglesey was by far the most talked-about construction project of the late 1840s. It showed the achievement that was possible even though there was little theory about and virtually no experience with such a structure. The immensity of the task of fabricating nearly-five-hundred-foot-long, 1,500-ton wrought-iron tubes and assembling them into a bridge drew considerable attention from technical and nontechnical onlookers alike.

The site of the Britannia Bridge project was frequently the scene of gatherings of engineers and others interested in design and construction, and Lucas's painting records only one such occasion. Joseph Paxton was attending a railway-board meeting there when he made his famous sketches for the Crystal Palace, and Isambard Kingdom Brunel no doubt gained from his visits to the Menai site much confidence for raising the spans of his own Saltash Bridge a few years later. Stephenson's bridge was the subject of some of the earliest photographs recording a construction project in progress, and books by Edwin Clark, the resident engineer, and William Fairbairn, whose model tests defined the geometry of the tubular girders, make the bridge one of the most documented engineering projects of the nineteenth century.

To today's eye, the configuration of the bridge looks utterly anomalous, without obvious structural antecedents or descendants, and the apparently functionless height of its towers above the tubes raises questions about Stephenson's engineering knowledge and intent. Still, although the structurally innovative qualities of the bridge are not immediately apparent and were overshadowed within five years of its completion, the conception and execution of the Britannia Bridge across the Menai Strait remains one of the most significant achievements not only of the mid-nineteenth century but also of all time. The story of the design and erection of this bridge remains relevant and instructive for understanding and practicing engineering today.

But why Robert Stephenson failed to apply a seemingly more logical arch or suspension principle at the site and chose instead to build his bridge out of longer, heavier, and more complex beams than had ever been used before, and why he thus brought upon himself engi-

Britannia Bridge

neering problems of an unprecedented magnitude, are questions with answers that lie in social and political as well as technical considerations. The questions and answers are understandable only in a historical context.

Among the masterpieces of bridge building in the first half of the nineteenth century is the suspension bridge across the Menai Strait designed by Thomas Telford and completed in 1824. In this 580-foot span high above the water, Telford used all three contemporary bridge-building materials: stone, wood, and iron. The towers and arched approaches were of stone, the suspended roadway was of wood, and the suspension chains were of wrought iron, which was known to be much stronger and more reliable in tension than the cast iron used in the compressed arches of iron bridges. Telford made sure he understood the behavior and strength of the iron by testing it.

In the 1840s, when it came time to bridge the Menai Strait with a railroad, the use of Telford's existing suspension bridge came immediately to mind. However, since its roadway would deflect excessively under heavy steam locomotives, driving the trains of the Chester & Holyhead Railway over the bridge was out of the question. George Stephenson, the great railway engineer and Robert's father, suggested that the railroad cars be unhitched from their locomotive when they reached the bridge, hauled over by horsepower, and then rehitched to another locomotive on the other side. Such a procedure was not in keeping with the image of a railroad that was intended to speed the mail between London and the ferry to Dublin, however, and alternatives to bridging the strait were sought.

Since the tides are particularly tricky in the Menai Strait, making it difficult enough to navigate in its natural state, the Admiralty did not want any obstacles to shipping placed in the water. This eliminated the possibility of erecting falsework or piers anywhere but possibly on the Britannia Rock, about a mile south of Telford's suspension bridge. The rock is about five hundred feet from either shore, so it was natural to design a bridge supported there from beneath by arches or from above by a tower and cables. The restrictions imposed by the Admiralty had driven Telford to choose a suspension design earlier in the century, but now that the suspension principle had been dismissed as unsuitable for a railroad—not because of lack of strength but because of lack of stiffness and reliability in high winds—an alternative to conventional bridge designs or construction methods had to be found. Robert Stephenson even developed a scheme whereby an arch would be entirely assembled near the shore on wooden scaffolding (known as centering) supported on pontoons, then the whole assembly floated into place at high tide so as to insert the arch between the abutments with a minimum of interference with shipping. This plan posed its own problems. Even if an arch could be erected without any centering at all—a scheme that had been proposed earlier by Telford and was to be executed years later by James B. Eads at St. Louis—the shoreward clearance of the arch over high water could not be achieved without lifting the railroad well above the prevailing land elevation.

Faced with the problem of bridging the Menai Strait without arches and without falsework, Robert Stephenson looked elsewhere for solutions. It was natural for him to come to regard iron as the primary material of his bridge, for that had been the direction in which bridge engineering was inexorably being driven in the nineteenth century. Furthermore, given all the social, political, and technological restrictions placed on the bridge, Stephenson believed iron could be used in a structurally efficient way to obtain the necessary strength and stiffness. The understanding that economic considerations include matters beyond initial cost remains central to engineering today: In the middle of the nineteenth century, iron was not the least expensive material, but its cost had to be weighed against the factors of fire and rot resistance. Although the price of iron had been dropping dramatically, it was actually remarkably high in 1846, when contracts for Stephenson's bridge were let—fifteen pounds sterling per ton for plate iron. When the cost of assembling the required 4,680 tons of plate into bridge girders was figured in, the outlay for the ironwork was well over half the total expense of the bridge, which was to amount to six hundred thousand pounds when it was completed in 1850.

Nevertheless, to the board of directors of the Chester & Holyhead Railway Company, Stephenson's scheme appeared to be the most economical means of achieving the desired results within the constraints of the time. His idea of using deep wrought-iron girders to stiffen the deck evolved into one employing a large, hollow tube through which the trains would pass, and Stephenson imagined using suspension chains to supplement the tube, not only to provide support during construction but also to support the great weight of the tubular structure after it was assembled. But as Stephenson thought more about the arrangement of the tube, its structural behavior became clearer in his mind: "The top and bottom plates performed precisely the same duties as those of the top and bottom webs of a common cast-iron girder. It was now that I began to regard the tubular platform as a beam, and that the chains should be looked upon as auxiliaries."

But what worked in Stephenson's mind and in his rough sketches had next to be put in precise quantitative terms. How thick should the iron be? How close should the rivets be driven? Exactly how should the top and bottom of the tube be constructed? Indeed, Stephenson realized that before he could begin to answer these questions he had first to deal with a more fundamental matter: Should the tubes be rectangular, oval, or circular? The answers were not easy, for although cast-iron girders fifty, sixty, and seventy feet long were not uncommon on existing railway lines in 1845, Edwin Clark noted that "the scientific principles of construction of such girders were not at once recognized or learned, and we consequently find excess of iron in most instances, and mistaken construction in others."

The technical problem was that, although the behavior of solid beams could be easily demonstrated, explained, and even calculated, the behavior of a hollow tube five hundred feet long, thirty feet deep, and weighing 1,500 tons raised all kinds of uncertainties. The iron tube would behave generally like an I-beam, but Stephenson knew that there most likely would be unknown differences. The tubes that Stephenson envisioned could conceivably behave in ways that he could not imagine without doing some experiments.

When Stephenson consulted with William Fairbairn, who was then well-known for his investigations into the strength of cast iron, the bridge engineer was reassured that the idea of iron tubes would work. Fairbairn related stories, which Stephenson had also heard from a shipbuilder on the railway board, of iron ships hundreds of feet long that had survived being tossed about in the sea and even being accidentally supported with one end raised on a wharf. A great hollow beam supported in the air only by its ends would not be unlike a ship in such a predicament. However, ships were also not exactly like the tubes Stephenson had in mind, and so Fairbairn could not tell exactly how stiff or strong the bridge tubes should be.

Fairbairn became engaged to assist Stephenson in an "experimental inquiry." Although there was growing confidence among the engineers that the several sections of tubes would be strong enough to support themselves, in order to get a bill authorizing the construction of the bridge through committee in the House of Commons, Stephenson had to "leave the impression upon the minds of the Committee that at all events the chains might be left as auxiliaries to the tube if necessary." Thus, the masonry towers, which would be built before the tubes could be jacked into place, had to be tall enough to receive suspension chains, if necessary.

In the meantime, Fairbairn tested models of tubes of various shapes, supported by their ends and loaded in the middle. The experiments showed clearly how and under what weight a riveted tube would break, and they led Stephenson to realize that rectangular tubes were best for the purpose. In failing, the models also identified the weak spots in the structure. These were beefed up in later experiments, and so the final configuration of iron in the tube could be established. Stephenson and Fairbairn did not seem especially inclined to determine which parts of their broken model tubes were overly strong, and so they made no systematic attempts to trim excess material where it was not needed. Had they done so, the final design of the Britannia Bridge might have been more economical. But such economy would have been gained at the expense of time and so might have been false economy to the board of directors, who wanted to see the rail link completed as soon as possible. As it turned out, the largest tubes Stephenson used were 460 feet long and thirty feet deep at midspan, a ratio of length to depth of about fifteen to one. Engineers today would characterize the tubes as slender structures.

Even with Fairbairn's methodical experiments, it would have been risky to extrapolate from the experiments alone the exact dimensions of the cross section of a full-size tube. A quantitative theory of the strength of tubes was necessary, and when Fairbairn suggested that Eaton Hodgkinson might help, Stephenson immediately consented to engaging him, already "being familiar with the valuable contributions of this gentleman to engineering science." What Hodgkinson did was to express analytically the strength of a wrought-iron tube in terms of its dimensions and an empirical factor determined from the results of Fairbairn's experiments. With the complementary support of theory and experiment, then, the detailed design of the tubes could be determined and construction could commence. The combined experimental, empirical, and analytical approach to engineering remains to this day a paradigm for designing large and complex systems.

The tubes were to be assembled on wooden platforms beside the water and then floated into place between the bridge piers, the construction of which had long since begun. Because of the continuing uncertainty as to whether or not chains would be required—Fairbairn maintaining that they would not and Hodgkinson that they would—the towers had been made tall enough to support chains to satisfy not only the House of Commons committee but also conservative engineering principles.

Since Stephenson had at the same time also been designing a somewhat less ambitious tubular bridge to cross the estuary at Conwy, some experience had already been gained there when it was time to float the first tube of the Britannia Bridge into place for its raising. However, the much greater height and tricky tides in the Menai Strait made this operation full of fresh uncertainty. Once the tube had been floated off the site of its fabrication, it was to be guided into place by a plan written out by Stephenson that was reminiscent in its order of the raising of the Vatican obelisk:

> All [four] capstans were fully manned by eleven intelligent superintendents, four hundred and fifty labourers, sixty-five sailors, and twelve carpenters. Each capstan had forty-eight men. The number of hands in each set of pontoons was one hundred and five, and six boats, with crews and spare line, attended the floating-tube in its progress. Two steamers were kept in readiness in case their services should be required.

Thousands of onlookers had gathered on the shore to watch, but they were disappointed when one of the capstans gave way under the strain and the tube had to be returned to its place of assembly. After operations recommenced, there were further difficulties, as one of the eight-inch towing lines snapped, and then a twelve-inch line on another capstan got snagged and could not be payed out properly. This capstan was dragged away from its platform, and had something not been done the tube might have floated away or the pontoons might have been crushed upon the rocks. According to Clark's account:

> The men at the capstan were all knocked down, and some of them thrown into the water. In this dilemma Mr. Charles Rolfe, who had charge of the capstan, with great presence of mind, called the visitors on the shore to his assistance, and, handing out the spare coil of the 12-inch line into the field at the back of the capstan, it was carried with great rapidity up the field, and a crowd of people, men, women, and children, holding on to this huge cable, arrested the progress of the tube.

Arresting 1,500 tons of iron was not accomplished without a great effort, which was the subject of a vivid account by the George and Robert Stephenson biographer L. T. C. Rolt:

> Never was there so dramatic a tug-of-war. It must have seemed to those who witnessed it that puny human muscles could not hope to regain control over the huge iron hulk which was swinging so relentlessly on the tide. Yet sheer weight of numbers prevailed. At first the struggling, shouting snake of humanity was dragged irresistibly forward towards the sea. Then it checked, held fast on the sea's edge. Finally, amid deafening cheers and cries of encouragement, slowly, slowly, fighting for each heel-hold, it began to move back. Obediently the great tube checked its career, drew back also and then began to swing majestically towards its appointed place. At last, with a thunderous reverberation, it struck the base of the Anglesey pier and the battle was over.

When the first tube was finally floated into place between the piers and began to be hoisted up by hydraulic jacks, it became indubitably clear that it was strong enough to support itself without chains. However, the masonry already having been erected well above the level of the tubes, it remained an artifact of uncertain design.

When on March 5, 1850, Robert Stephenson rode on the first train to pass through the Britannia Bridge, it was declared an unquestionable structural success. Furthermore, Edwin Clark expected the environmentally disruptive scars of years of construction eventually to disappear and the bridge to become fully integrated into the larger landscape and culture of the region. He wrote:

> It is, indeed, a source of high gratification to have been instrumental in realizing so magnificent a conception of a master mind; and it is a pardonable vanity to believe that such a record of the industry of an intelligent people will increase in interest when all traces of this once busy scene shall be effaced; when the noise of the hammer has ceased, and the ephemeral village of the builders has disappeared; when the fern and the moss shall have ventured to invade even the massive pile that has displaced them; when the sea-weed with its coral inhabitants shall have mantled the foundations, and the wild bird, so long afrighted, shall return to its solitude, and rear its progeny beneath the marble shelter of the towers.

It is ironic that Clark wrote so movingly of the bridge's hospitality to the local flora and fauna, for almost as soon as the Britannia Bridge was opened to traffic, it became clear that it was inhospitable to human travelers. After all, a 1,500-foot-long tunnel in the sky collected a lot of smoke and soot as coal-burning locomotives passed through it. Furthermore, passengers came to dread the Britannia Bridge not only because it was a dirty experience but also because it was a very hot one. In the midday summer sun, the temperature inside the wrought-iron tubes became unbearable.

Within a few short years, a couple of tubular bridges were built in Egypt, and one was built across the St. Lawrence River at Montreal, but even the slotted top of that mile-long sky tunnel did not remove the objections to the smoke and heat. Yet had tubular bridges been the only structural solution, railway passengers might have endured many more of them. After all, the Britannia Bridge did remain in service until 1970, when a fire was accidentally set in the timber roof that had been erected over the tubes to protect them from corrosion. The heat of the fire left the tubes warped and deformed beyond repair, but now that tall ships no longer ply the strait, the bridge could be rebuilt as an arch.

What made the Britannia Bridge a dinosaur even in the 1850s, however, was its cost. When in 1845 there appeared to be no viable alternative to Stephenson's scheme, it was considered an economical design. While Stephenson and his colleagues were designing and erecting the great rectangular tubular bridge, however, perhaps with the tunnel vision necessary to see it through to completion, other engineers were planning and designing other kinds of bridges. Since these other engineers were not faced with nor so committed to the task of making the tube concept work, they were free to explore alternatives.

A dramatic challenge to the concept of the Britannia Bridge was to occur within five years in America. John Roebling, manufacturer of wire rope and erector of canal and carriage-road suspension bridges, successfully erected a railroad-carrying suspension bridge with an 821-foot span between towers. Not only was the deep timber bridge deck that he employed capable of carrying fully loaded railroad trains with very little deflection of the trackbed, but the bridge vibrated even less under the iron horse than it did under the hoofbeats of the real horses pulling wagons on the lower carriageway. Roebling's Niagara Gorge Suspension Bridge proved once and for all that a suspension bridge, which Stephenson effectively rejected, could carry railroad trains and withstand high winds—and do so very economically.

What also showed the design of the Britannia Bridge to be an uneconomical use of materials was Isambard Kingdom Brunel's Saltash Bridge. The Tamar River crossing at Saltash presented the same structural challenge to Brunel as the Menai Strait did to Stephenson: The thousand-foot-wide crossing could be broken by only a single pier in the water, and the Admiralty again demanded its headroom. Even though Brunel's design used materials efficiently, what was feasible to one railway company at one time was not necessarily so to another at another time. Brunel actually began preliminary construction trials on the Saltash Bridge in 1848 and had demonstrated the practicality of his scheme, but lack of capital forced the Cornwall Railway Company to suspend the project for three years. When the company decided to proceed, it was with a less costly bridge, on a line that was to have a single rather than a double track predominating. According to one of Brunel's reports, a savings "of at least 100,000 pounds sterling" could be realized by adjusting the design of the railroad across the Tamar. But the bridge was still a clear success, and it was opened as the Royal Albert Bridge in 1859. (Informally, it is known as Brunel's bridge, a designation reinforced by the fact that its entrance portals bear the inscription "I. K. BRUNEL, ENGINEER.")

Such is the relative meaning of economy in large engineering projects. What seems wasteful in retrospect may have seemed the economical thing to do at the time when hard decisions had to be made—not necessarily by engineers alone but in concert with the businessmen and politicians who were fiscally and socially responsible for the project. When the board of the Chester & Holyhead Railway or members of Parliament wished to postpone no longer the completion of the rail link from London, whether to increase railroad revenue or to speed the mail to Dublin, the die was cast to take the known economic and engineering risks, even with a paucity of engineering and scientific knowledge, rather than wait for more economical or structurally superior developments in an uncertain future.

Indeed, the economic and political decision to build the Britannia Bridge, with all its inherent uncertainties, might actually have brought the improvements of Roebling's Niagara bridge and Brunel's Saltash Bridge about faster. Had the Britannia never been started or completed, it could not have served as a standard against which to measure improvements. Without the successful completion of the Britannia Bridge, the financiers of the later bridges might not have had the temerity to proceed as early as they did.

Tower Bridge

Tower Bridge is to London what the Golden Gate Bridge is to San Francisco. Tourists flock to these instantly recognizable landmarks, and their images are ubiquitous in the souvenir shops of their respective cities, appearing on everything from T-shirts to teaspoons. Although many West Coast tourists can identify the Golden Gate as a suspension bridge—indeed, as one of the largest and grandest examples of the form—few visitors to London can properly categorize Tower Bridge. In fact, many nonspecialist engineers have a difficult time classifying the unique Victorian structure, for it is one of the most unusual bridges in the world. Like many unusual artifacts, its form arose from the unique conditions under which it was designed and built.

During the mid-nineteenth century, London Bridge, just upstream from where Tower Bridge now stands, was the first barrier that tall-masted oceangoing ships encountered in their journey up the Thames. Some ships anchored immediately downstream of the old bridge, in the deeper midstream water known as the pool. In that location, where the banks of the river were low, ships were loaded and unloaded by smaller, shallow-draft vessels known as lighters. The Tower of London stood then, as it does now, beside that part of the river, on naturally rising ground, and immediately upstream from St. Katharine's Dock.

A need for additional river crossings developed with the growth of

Greater London during the latter half of the nineteenth century. London Bridge, always a bottleneck, had become unbearably crowded, and so a new span was proposed to be constructed a little way downriver. Financing for the project was assured through the well-endowed Bridge House Estates Trust, which had its origins in the thirteenth century, when tolls were collected on London Bridge and set aside for its maintenance. The ample fund came to be employed for transportation improvements beyond London Bridge proper.

Designing a new bridge across the Thames in the latter nineteenth century was a formidable task, not so much for structural reasons but because of the historical, topographical, and commercial constraints imposed by the location. The north-side approaches to the new bridge were placed most naturally between the Tower of London, the bridge's namesake, and St. Katharine's Dock, where the banks of the river are low. The bridge either had to be a high-level crossing, so as not to obstruct shipping, or have a movable span. A bridge with high clearance would have necessitated long approaches, which would have added to the expense, not to mention presented an unwelcome structure beside the Tower of London. By incorporating a movable span in the center of the bridge, the side spans could be built close to the water and hence minimize the approach viaducts.

Constraints clearly affect the nature of bridge design as they do all of engineering. Since they must be taken into account, usually through compromise, few bridges can be first sketched on paper the way they are finally built, and Tower Bridge was not one of the exceptions. Among the early designs was a low-level crossing of the kind once proposed across the Delaware River at Philadelphia: one with a divided roadway to eliminate the backup of traffic that accompanies the operation of a movable span. It was not difficult to foresee the problems associated with controlling a ship in the narrow confines between the forks of such a river crossing, however, and so other solutions were sought.

Tower Bridge as it exists was designed in the 1880s by the engineer John Wolfe Barry, who worked on various dock and pier projects around Britain, including the Southend Pier at the mouth of the Thames and the Barry Docks near Cardiff, which enabled the Barry Railway Company to capture some of the export trade in coal from south Wales. Barry was to be knighted and serve as president of the

Institution of Civil Engineers shortly after the completion of Tower Bridge, but such honors would come only after much criticism.

Having received the requisite parliamentary approval, construction of Tower Bridge began in 1886 with the establishment of midriver foundations. Because it was imperative that no significant settlement develop when the superstructure was erected and later loaded with traffic, preliminary work for the structure included determining how much load could be borne by the London clay in the riverbed. To do this, a test cylinder was sunk and loaded until it began to settle, which started at 6.5 tons per square foot. A conservative maximum load of four tons per square foot was then decided on, after which the size of the foundations was fixed according to the design load of the superstructure, by then established on paper. The size of these foundations generated much discussion in the engineering literature of the time, for only those of the Brooklyn Bridge, completed just a few years earlier, were larger.

It was next necessary to excavate the riverbed to where the total load could be supported, and that required the construction of caissons and cofferdams to keep the water out while the excavation proceeded. The concept of a cofferdam—consisting of a ring of closely fitted piles—dates from Roman times. Pneumatic caissons, which were used to dig and ultimately contain the foundations of the Brooklyn Bridge, are essentially large, open boxes sunk upside down in a riverbed, thereby creating a plenum of air within which workers excavate material as the caisson sinks to the desired depth. The caissons used for Tower Bridge were open to the atmosphere and thus did not present the dangers of a compressed-air environment.

For establishing each midstream foundation of Tower Bridge, a group of eight twenty-eight-foot-square caissons was employed, along with four triangular caissons to give the characteristic pointed shape to the ends of the piers. To enable the caissons to be leveled individually as they sunk, they were spaced a couple of feet apart, with piling used between them to form a cofferdam. When the proper depth was reached, the caissons were filled with concrete, upon which brickwork was erected, which in turn served as a base for the steel and masonry construction above the waterline. In reporting on the construction project, the American trade journal *Engineering News* noted that the caissons were made of steel. That was necessary because the use of timber caissons was specifically forbidden by Parliament, presumably because of "the great cost of suitable timber in England."

The journal further noted that as the river muck was excavated and the caissons descended into the river, the space filled up with water after a high tide. That necessitated pumping out the water, leaving only "two to six hours of dry work between successive high tides." According to *Engineering News,* "this awkward arrangement seems to have been the result of an error" in sizing the various steel parts of the cofferdam, and the journal expressed surprise that the error was not corrected as the work proceeded. Presumably the inconvenience was tolerated because the cost of pumping out the water after each high tide was determined to be less than that of redesigning the ill-sized steelwork. The journal further criticized the fact that it took more than four years to finish building the two main piers—"certainly very slow work"—and cited as exemplary the rapid progress for several contemporary American projects, including bridges at Memphis, Tennessee, and Poughkeepsie, New York.

The substructure of most bridges is usually forgotten once the superstructure is erected, and Tower Bridge was no exception. The bridge proper, which has been described as "a double-leaf bascule with suspended side-spans, incorporating two high-level footbridges," began with the erection of steel for the towers. The steelwork framing made the towers effectively freestanding four-story buildings, which were to be covered with rough-faced granite to give them an appearance in keeping with the character of the nearby Tower of London.

The external appearance of the bridge towers was designed by the architect Horace Jones, who devised a Victorian Gothic design that was and has continued to be the object of much criticism. (Indeed, the matter of enclosing the load-bearing steel frame within a masonry facade foreshadowed by some decades a similar intention for the George Washington Bridge, which cladding of course remained unrealized.) Among the outspoken late-Victorian critics of the stone-over-steel towers of Tower Bridge was H. Heathcote Statham, a fellow of the Royal Institute of British Architects, who wrote in his paper on "The Architectural Element in Engineering Works":

> The Tower Bridge . . . represents the vice of tawdriness and pretentiousness, and of falsification of the actual facts of the structure. It is

Tower Bridge

stated that the exterior clothing was designed by an architect; he cannot have been a very eminent one, as we never hear his name; it looks to me more like what results from the advertisement we sometimes see—'Wanted immediately a draftsman; must be an expert Gothic hand'—it is draftsman's architecture. The exceedingly heavy suspension chains are made to appear to hang on an ornamental stone structure which they would in reality drag down, and the side walls of the apparently solid tower rest on part of the iron structure, and you could see under them before the roadway was made up. All architects would have much preferred the plain steel structure to this kind of sham.

J. A. L. Waddell approvingly quoted this passage in the chapter on aesthetics in design in his massive 1916 treatise, *Bridge Engineering.* He noted further that "any American engineer travelling in England would do well to visit the bridge so as to see for himself how far in that country extravagance in design can be carried and to what an extent the important factor of efficiency can be ignored." Waddell himself was open to criticism, however, for having perhaps too much concern for efficiency in his innovative but spare Halsted Street vertical-lift bridge over the Chicago River, an exact contemporary of Tower Bridge. But criticism after the fact is easy if one forgets the severe physical and political constraints under which most bridges are designed and built.

Each of the "chains" pulling on the stone towers of London's Tower Bridge is in fact two massive, rigidly riveted, crescent-shaped trusses hinged together. The suspended spans are each about 270 feet long, a modest distance compared to the cable-supported 1,595-foot central span of the contemporary Brooklyn Bridge. In fact, the truss-suspension system of Tower Bridge may have reflected a long-standing British concern with wind effects on bridges. In addition to the memory of how in the 1830s the wind tossed about the suspended decks of the Menai Bridge and Brighton Pier, there was the more recent experience with the high girders of the Tay Bridge, which were believed to have been blown off their piers and into the water. The Forth Bridge, under construction when Tower Bridge was designed, addressed the matter of the wind with massively braced cantilever arms reaching out from great steel towers 1,710 feet apart. The truss chains of Tower Bridge were thus in keeping with the spirit of the times in Britain. (In fact, many a famous near-contemporary bridge, including Pittsburgh's Point Bridge, completed in 1877 over the Monongahela River, and Gustav Lindenthal's late 1880s proposal for a three-thousand-foot suspension bridge across the Hudson River, incorporated similarly stiff suspension systems in preference to cables or even true chains.)

Although Tower Bridge was not a structure of remarkably long fixed spans, it was notable for its movable center span. The idea of a drawbridge was consistent with the medieval-castle theme of the location, but a conventional draw span that was lifted by chains pulled by brute force would not do across the two-hundred-foot shipping channel mandated by Parliament. A bascule bridge, by contrast, incorporates a movable deck that is counterweighted, requiring relatively little power to raise and lower it.

The bascule design presented an alternative in crowded cities to the tall and generally unattractive lift-bridge concept favored by Waddell.

Indeed, a bascule design could so conceal the massive counterweights and machinery required for its operation that the movable nature of the bridge might hardly be noticed when it was not open. For example, the Arlington Memorial Bridge, completed in 1932 between Arlington Cemetery and the Lincoln Memorial and considered to be one of the most beautiful bridges in Washington, D.C., has a well-concealed steel

Arlington Memorial Bridge bascule span

bascule span between its graceful stone-faced concrete arches. In late-Victorian London, however, a bascule on the scale of Tower Bridge was a new concept and one that held much interest for engineers, almost more so for its machinery than for its structure.

The need to obviate any uneven settlement, so that the hundred-foot-long and fifty-foot-wide roadway of the thousand-ton bascule leaves would remain aligned for smooth operation, was among the prime reasons that substantial foundations were required for Tower Bridge. The power for moving the leaves was initially provided by two

360-horsepower steam-pumping engines, which pressurized water in accumulators to 850 pounds per square inch. That water pressure then drove rotary hydraulic engines, which in turn activated gearing that raised and lowered the bascule leaves. All machinery was installed in duplicate so that the breakdown of any one part would not jeopardize the efficient operation of the bridge. It is said that the only time the system did not work perfectly was the first time the machinery was tried, although reportedly on at least one hot summer day, in 1968, the bridge jammed because of the heat.

The principal power source was in the Surreyside (south) abutment of the bridge, and the water pressure was conveyed to the Middlesex (north) pier via piping that was located in the elevated walkways. It is unlikely that this pressure conduit was the primary reason for including the high-level girders between the bridge towers, for duplicate machinery could have been located on either side of the bridge. Rather, the elevated walkways, which were erected cantilever fashion out from the steelwork of the towers because Parliament did not allow scaffolding to obstruct the river during construction, could conceal a structural tie between the towers, which were pulled in the opposite direction by the suspension trusses. Not incidentally, the elevated walkways would provide a means for foot traffic even when the bascule section was raised. Tower Bridge was officially opened by the Prince of Wales on June 30, 1894, and it was soon carrying seventy thousand pedestrians and eight thousand trams daily. The bascules were raised as many as twenty times per day to allow ships to pass.

Over the years, with a decline in use of the nearby docks and in the number of tall ships wanting to pass Tower Bridge, it came to be opened less and less frequently. Now it is a rare occasion on which the bascules are raised, which does not disappoint the great number of car, taxi, bus, and lorry drivers who use the bridge daily. The elevated walkways, which had become less necessary with the decline in traffic interruptions and which had been closed to discourage suicides from the 140-foot height, were enclosed in glass and reopened in the 1980s as part of the museum devoted to itself that the bridge and its towers now house. The seldom-used opening machinery has been converted to electrical power, but some of the original steam-driven equipment is preserved nicely in the Surreyside abutment as part of the museum.

For many decades, Tower Bridge remained the first structure across the water that a ship sailing up the Thames would encounter. Since 1991, however, that distinction has been held by a new cable-stayed bridge at Dartford, some twenty miles downstream. Officially known as the Queen Elizabeth II Bridge, this high-level crossing carries southbound outer-ring-road traffic, thereby relieving the congestion that had choked a tunnel at that location, freeing it to carry only northbound traffic. The first potentially true obstacle to shipping is the Thames Barrier. This unique flood-control structure, completed in 1984 near Greenwich, comprises a series of ten movable gates, each two hundred feet wide, across the entire width of the river. The individual gates, which can be rotated into recesses in the river bottom to allow ships to pass, can also be raised by hydraulic machinery to stop surge tides from the North Sea that would otherwise threaten the populous city upstream. But no matter how many barriers and bridges may be thrown across the River Thames, and because of the unusual nature of its engineering, the structurally odd Tower Bridge will no doubt remain the most recognizable and popular.

Drawing Bridges

Many a person who has driven through the Middle Atlantic states on Interstate 95 has dreaded the approach to Washington, D.C., where traffic often slows to a crawl, if not to a complete stop. Heading north through Virginia toward Maryland, I-95 widens from two to three and then to four and five lanes and more, to accommodate the increased traffic volume and interchange complexity as one nears the capital. About twelve miles south of Washington, traffic invariably becomes more erratic and frenetic, as cars and trucks jockey for position in anticipation of the branching of the highway into I-395 and I-495, the former going through Arlington into the District of Columbia proper and the latter being the Outer Loop of the Capital Beltway, which bypasses Washington—but not its traffic. The westward leg of I-495 carries traffic headed for Fairfax, Virginia, and Bethesda and Rockville, Maryland. Those traveling I-95 north to Baltimore and on to Philadelphia and New York generally continue on I-95, which coincides with the eastward leg of I-495.

About six miles after making the practically irrevocable decision to head eastward on I-95/I-495 in order to continue north, drivers find that the interstate highway crosses the Potomac River on the Woodrow Wilson Memorial Bridge, the only federally owned bridge in the entire interstate-highway system. Since by Coast Guard regulations the drawbridge may not open during periods when it is carrying a heavy volume of vehicle traffic, most highway travelers and commuters never see the

80

bridge in action, but in the late 1990s it was opening about 220 times a year nonetheless. This, along with the sheer volume of traffic carried by the bridge—two hundred thousand vehicles per day, almost three times what it was intended to carry—make it a location that the American Automobile Association has described as "one of the nation's top ten bottlenecks." It is indeed that, as well as obsolete, and why it might be so becomes clear when one looks at the history of the bridge.

Design and construction of the Wilson bridge was authorized in 1954 to link the Maryland and Virginia portions of what was then known as the Washington Circumferential Highway, and construction began in 1959. This means, of course, that the design standards followed were those of the 1950s, which did not take fully into account the volume or intensity of traffic contributed by today's vehicles or the resultant problem of structural fatigue—a phenomenon by which the cumulative effect of loads of even moderate intensity repeated millions of times produces damage that the less frequent application of those same loads would not. Loads of higher intensity, such as those imposed by trucks now heavier than those anticipated in 1950s standards, accelerate the onset of fatigue.

The Wilson bridge was designed with three lanes in each direction, which must have seemed more than adequate at the time. The intended capacity of the bridge—seventy-five thousand vehicles per day—also must have appeared quite reasonable, since the actual traffic load was only nineteen thousand vehicles per day. Bridges encourage traffic, however, and by the late 1960s automobiles and trucks using the Wilson bridge had exceeded its design volume. Approach roads were widened to three and four lanes in each direction, but the bridge remained at three each way because of structural limitations. Furthermore, in the mid-1970s it was decided to abandon plans to extend I-95 through Washington proper, and that leg of the interstate system was permanently redirected to its present route along the I-495 Capital Beltway, which includes the Wilson bridge. By the late 1970s, when structural deficiencies in the bridge were identified, a major rehabilitation project was begun, with much of the deck-replacement work taking place between 10:00 p.m. and 5:00 a.m. so that the full six lanes of the bridge could remain open during most of the peak traffic hours and accommodate the 108,000 vehicles that were then using the span each day.

A major study was initiated in the late 1980s to consider replacing the bridge and reconstructing the roads in its vicinity, and a concept competition was held to find innovative solutions to the problem. The consensus solution, a fourteen-lane bridge to the south of the present Wilson bridge, was not found fully acceptable by a jury of experts, but an environmental-impact study of the inevitable project was carried forward nonetheless. By the early 1990s, traffic volume on the bridge was more than double what it was designed to carry, and in 1992 a coordination committee representing those most affected by the project was appointed by the Federal Highway Administration to oversee a new improvement study for traffic problems in the area. Specifically, the committee was charged with identifying a "solution which enhances mobility while assuring that community and environmental concerns are addressed." Among the outcomes of the study was the finding by a bridge-inspection firm that the useful life of the Wilson bridge would be reached in 2004, a limit determined principally by the accelerated deterioration taking place under increased truck traffic, amounting to about 17,500 trucks per day carrying 1.3 percent of the total volume of all U.S. truck shipments by manufacturers, wholesalers, and others.

Among the options considered for improving the flow of traffic was to build a new span ten miles downstream. However, since 85 percent of the traffic over the existing bridge has its origin or destination within the Washington metropolitan area, it was estimated that the new bridge would reduce traffic across the present span by only 10 percent. The problem of replacing Wilson bridge with a larger-capacity span at the same location also involved redesigning nearby highway interchanges, which were overburdened as well. Such redesign and construction would have obvious impacts on the local community. Public meetings and citizen workshops provided opportunities for input from interested groups.

In 1996, the coordination committee, comprising transportation and elected officials from the affected area, decided upon a preferred conceptual design for a replacement bridge: a pair of six-lane spans, the exact nature of which would be determined via a design competition. While such a competition might account for only 10 percent of the total final design effort, it would provide a basis into which to incorporate subsequent refinements and the many detailed design decisions that go into such a project.

Although an unusual means of selecting a bridge design in the United States, design competitions have a long and rich tradition elsewhere in the world, especially in Europe. In the case of procuring a design for the replacement of Wilson bridge, the method was chosen at least in part because of a highly successful application of the design-competition concept about a decade earlier to replace a state-highway bridge in the historically sensitive region of Annapolis, Maryland.

In the late 1980s, the sixty-year-old drawbridge carrying Maryland Route 450 over the Severn River at the state capital was badly in need of replacement. The bridge site is at the eastern gateway to the city of Annapolis, and the western approach of the bridge is on the grounds of the U.S. Naval Academy. The city is of historic significance and contains numerous fine examples of Georgian and Victorian architecture. The bridge site is within view of residential communities located on the bluffs above the waterway, and it is not far from the mouth of the Severn River, which empties into scenic Chesapeake Bay, a body of water often crowded with recreational sailboat traffic. In short, a new bridge across the Severn would be seen in many different ways by many different groups, and the challenge to engineers was "to develop a design for a bridge which will respect, enhance, and encompass these components."

It was decided to conduct the first international engineering-design competition held in the United States in almost a century, in order "to bring to this important site those people who know the most about bridge engineering, and to encourage them to think technically, economically and aesthetically about how best to bridge it." The language echoes that found in David Billington's *The Tower and the Bridge: The New Art of Structural Engineering,* which emphasizes "efficiency, economy, and elegance" in bridge building. Billington, a professor of civil engineering at Princeton University who has served as a consultant to the Maryland Department of Transportation on bridge aesthetics, has long been a strong advocate of design competitions. His ideas no doubt had a major influence on the decision to seek a new bridge design through such means. It was believed that the resulting bridge would be "a work of structural art which will make all of Maryland proud." As firm evidence that the bridge should be considered as more than just a state-highway project, the competition was cosponsored by the Maryland State Highway Administration and the Governor's Office of Art and Culture.

The program and rules of the Severn River Bridge Design Competition were developed during a project-planning study, which included a public hearing. The program and rules were issued early in 1989, and advertisements announcing the competition were placed in local newspapers and in the engineering press, including *Engineering News-Record*. The new bridge was to be a fixed, high-level crossing, thus eliminating traffic delays caused by the raising of the old drawbridge. Also, the new structure was to be constructed immediately south of the existing bridge, which would be removed once the new one was in service. The number and width of traffic lanes, shoulders, and sidewalks were specified, as was the clearance for shipping (about seventy feet over a channel width of 140 feet) and the minimum main-channel span (three hundred feet). The cost of the new bridge was estimated to be between twenty-five and thirty million dollars, and prospective competition entrants were advised that designs with higher estimated costs would be evaluated but penalized on economic grounds.

Evaluation criteria for the judging of entries were encapsulated into the following statement, which appeared in the program and rules: "This is an engineering design competition, meant to integrate the best state of the art thinking about economy, aesthetics and technology in a single structure. High quality in one factor must be matched by high quality in the other two. The bridge as a work of structural art must enhance its environment."

The process for selecting a winner of the competition occurred in two stages. In the first stage, "any interested qualified party" could submit a letter of interest and qualifications, a list of joint-venture constituents, proposed subconsultants and subcontractors, and the résumés of the lead and associate bridge designers, along with a description of at least three of their bridges. This information was to serve as the basis for choosing a maximum of six finalists, who in the second stage of the competition would each prepare a preliminary design for the bridge. The completion of this stage of the competition with an acceptable design would secure the finalist twenty thousand dollars toward the cost of services. An additional twenty-five thousand dollars was reserved for prize money.

Twenty-one letters of interest were submitted, from engineering firms throughout the United States and from a few engineers in Europe, and the selection committee chose six of them to be finalists. These included five American firms and the Swiss firm of Calatrava Valls, all of which were required to submit a preliminary design within four months. More than just conceptual, these preliminary designs were to contain sufficient detail so that the sizes of major structural components of the bridge could be discussed and realistic cost estimates could be established. In the end, the Swiss firm did not submit a design, but the five American firms did, and these were kept anonymous by identifying them only as Entrants B through F. The fourteen-member selection jury included David Billington and Christian Menn, the Swiss bridge engineer whose own designs are works of art.

Among the jury's major criteria for selecting a winning entry were "effectiveness in integrating aesthetics, technology and economy"; "effectiveness in meeting the visual, symbolic, historical, and functional goals of the site"; and "constructability, maintainability, and likely durability" of the design. The jury found that all five entrants met the requirements of the program and rules, and it chose for first prize and the design award the entry from the Towson, Maryland, firm of Greiner, which specified twin steel trapezoidal box girders with a profile over the main span piers shaped into a graceful haunch or deepening of the girder. Among the reasons the jury gave for choosing this design was that "the shape of the girder in profile and the cross section of the superstructure combine to create a graceful reflection of the geometry of the roadway and the forces on the structure. The superstructure will appear as a thin, curving ribbon arching over the Severn River." The jury did have some suggestions for refinements in the design, however, including combining two awkward curves in the bridge into a single flowing curve, replacing the four columns under the haunches with two, and removing the pylons that were included at each end of the bridge to serve as "gateways to Annapolis," the jury feeling that "the bridge itself should be the gateway." The new Severn River Bridge has now been in service for several years, and it appears to be everything that the jury believed it would be.

It was the success of the competition that emboldened Maryland's Office of Bridge Development to repeat the process for the replacement of the Woodrow Wilson Memorial Bridge. The Maryland Department of Transportation, in conjunction with the Virginia Department of Transportation, the District of Columbia Department of Public Works, and the Federal Highway Administration, announced the new competition in 1998.

A memorandum of agreement, signed by the sponsoring agencies along with the National Park Service, the Advisory Council on Historic Preservation, and the state historic-preservation officers of Maryland, Virginia, and the District of Columbia, set down design goals for the new Wilson bridge. These began:

> 1. The Bridge (Potomac River Crossing) shall be a structure designed with high aesthetic values, deriving its form in relation to the monumental core of Washington, D.C., and shall be an asset to the Nation's capital and the surrounding region.
> 2. The concepts for the bridge shall be based on arches in the tradition of notable Potomac River bridges (e.g., Key Bridge, Memorial Bridge).
> 3. The bridge design shall employ span lengths which minimize the number of piers occurring in the viewshed of the Alexandria Historic District and other historic properties. Every effort shall be made to minimize the footprint of the project without adversely affecting safety and operations.

Nine further goals were specified, including the minimization of impacts on parklands, river views, and a cemetery in the area. Thus, in many ways, the Wilson bridge project was very similar to the Severn River one but on a much larger scale. The replacement bridge across the Potomac, including the reconstruction of four interchanges in the vicinity of the bridge, was described as "the biggest public works project in the Middle Atlantic states" and was expected to cost on the order of $1.8 billion, or sixty times the project at Annapolis. Unlike the new Severn River crossing, however, the new Wilson bridge would retain a draw span, in large part because local politics and the approach conditions did not allow for a sufficiently high-level bridge. The approach grade for such a bridge would have exceeded 4 percent, slowing trucks as they climbed the grade and thus hampering the smooth flow of traffic on the interstate. The new bridge will, nonetheless, provide a twenty-foot-higher clearance beneath the closed draw span, thereby reducing the estimated annual openings of the bridge from 220 to 65, or approximately one per week.

The selection process for the new Wilson bridge was to resemble the earlier design competition, but with the more or less autonomous jury replaced by a selection panel, which in turn was advised by four different committees—one each to address technical, constructability, historic, and citizens' issues. This last, especially, highlights the more highly politicized nature of the Wilson bridge project, and the establishment and continued operation of a Woodrow Wilson Bridge Center not far from the construction site in Alexandria emphasized the importance of public relations in dealing with nontechnical aspects of such a project. There was also a very effective and informative Web site.

The overarching objective of the competition was "to produce a fittingly world-class design" for the bridge, and seven design teams expressed an interest in pursuing that goal. Four teams were selected as finalists, and they were allowed to submit up to two entries each in the competition. There were seven entries, and they were considered by the advisory panels, whose input was taken into account by the selection committee, which was headed by Harry R. Hughes, former governor of Maryland and former secretary of the Maryland Department of Trans-

Design for Woodrow Wilson Bridge replacement

portation, and included Professor Billington. The design chosen was produced by the team of Steinman, Boynton, Gronquist & Birdsall in partnership with DeLeuw, Cather & Company, which are now both units of the Parsons Transportation Group.

Although the chosen design looks like an arch bridge, as required in the stated goals of the competition, it is not in fact a true arch. Light, open, curving, V-shaped concrete piers rise out of the water to support slightly haunched deck girders that combine with the lines of the piers to give the illusion of graceful arches spanning the river. In contrast to the fifty-seven piers of the present bridge, the new design has only eighteen, thus making for a light yet monumental structure. One jury member, a fine-arts commissioner, likened the V-shaped piers to the outstretched thumb and forefinger of "Neptune's hand reaching from the water to support the deck of the bridge." A citizen, upon viewing the design, noted that "the graceful, swooping lines of the repeated 'faux arches' had the appearance of several seagulls taking flight from the river."

Citizen relations were being actively developed along with the project, and the chosen design was unveiled to the public the day after the jury made its selection. The unveiling took place at the Woodrow Wilson Bridge Center on November 19, 1998, and a series of stakeholder-participation panels was convened the next month. These panels provided a forum for individual citizens and interested groups to voice concerns about various aspects of the larger project, in particular the highway interchanges closest to the bridge. In addition to representatives of local government bodies, the stakeholder panels included members of the Sierra Club, businesses, groups, the AAA, commuters, bicycle-user groups, and individuals representing persons with disabilities. The input of these panels was expected to identify potential problems with the design and construction of the interchanges and thereby allow modifications to be incorporated at the planning stages.

The new project promised to be a model of how to obtain a world-class design while addressing the needs and concerns of the people it is intended to serve. Its progress continued to be watched closely by engineers, planners, government officials, and citizens alike. They all had plenty of time to observe, for the construction schedule was projected to stretch over six years. The first work, installing the foundations for the new bridge, began in the fall of 2000. The first half of the new bridge—for the replacement bridge is in fact a pair of twin bridges— was originally planned for completion in 2004, and the whole project was scheduled to be completed in 2006. These milestones assumed, according to the project's Web site, "full project funding and no major unanticipated delays." Within five months of the unveiling of the design, however, a glitch did develop. Alexandria community activists who oppose such a large structure won in federal district court their case claiming that pollution effects of and alternatives to the twelve-lane design were not adequately considered. It was feared that the ruling could add six months to two years to the schedule. Another complication was the introduction in Congress of a bill that would cap the federal share of the cost of the span at nine hundred million dollars.

The cost of the bridge proper, which did not include the improvements to the interchanges on either side of the river, was estimated by engineers to be about five hundred million dollars, and so bids were naturally anticipated to come in around that amount. However, only a single bid was submitted, by a team of three construction companies, and it was for more than $850 million. Although some observers suspected collusion among the contractors, the single, high bid was attributed to the coincidence of bids being due at about the same time on several large construction projects, including the billion-dollar East Bay replacement span for the San Francisco–Oakland Bay Bridge, which naturally taxed the industry's capabilities. A further factor that was believed to have affected the bidding was the uncertainty until just before the bid deadline of whether union workers would have to be used. Maryland and Virginia had opposing views on unions, the former being pro-union and the latter having the reputation for being a right-to-work state.

The unexpectedly high bid was rejected by the Maryland Department of Transportation, which was administering the project because the bulk of the Potomac River—to the mean low-water line on the Virginia side—belongs to Maryland. In order to attract more and more reasonable bids, the project was broken up into smaller contracts, each of which was expected to be within the capabilities of construction companies that had shied away from the single large contract. The Bush administration's eventual refusal to provide federal funding if contractors had to abide by pro-union labor agreements also encouraged wider bidding.

In late 2002, multiple bids were received, and the contract for the bridge proper went to a partnership of the American Bridge Company of Coraopolis, Pennsylvania, and Edward Kraemer & Sons of Plain, Wisconsin, which together submitted the low bid of just over $185 million, only 11 percent over the estimate of the engineers at the Maryland State Highway Administration. The entire project, including dredging, foundation work, interchanges, and the appurtenances associated with any major highway improvement, was now expected to cost just under $2.5 billion. This amount was small compared to the $14 billion that Boston's Big Dig had then reached after numerous cost overruns, but the Wilson bridge project can be expected to experience its own overruns before it is completed.

The variety of setbacks has naturally delayed the projected completion date of the replacement Wilson bridge. The most optimistic estimate as of early 2004 was that the first six-lane span would be completed in 2006, at which time traffic will be shifted to the new bridge, the old bridge will be taken down, and work on the second six-lane span will begin. When it is fully completed, the structure will be turned over by the Federal Highway Administration to Maryland and Virginia, which will retain joint ownership. Regardless of the ultimate schedule and disposition of the bridge, however, in the interim many a traveler and commuter is likely to find increasing frustration in crossing the Potomac River on I-95, and many a driver on the Florida-to-Maine route can be expected to look toward the western leg of I-495 as an alternative way around Washington, or to I-395 and U.S. 50 through the heart of the District of Columbia, at least until the new Woodrow Wilson Memorial Bridge is completed in the yet-to-be-certain future.

An Eye-Opening Bridge

In early August 1996, a full-page advertisement appeared in the British magazine *New Civil Engineer* announcing a competition to design a footbridge across the River Tyne. The text of the ad, headed simply "a new bridge across the Tyne," was briefer than many classifieds but fully challenging: "We are looking for design teams who can create a stunning, but practical, river level crossing which fits this historic setting, opens for shipping and is good enough to win Millennium Commission funding. Teams who think they are up to the task should register for details of the design competition." The address given was that of the director of engineering services of the Gateshead Metropolitan Borough Council.

Gateshead is on the south side of the River Tyne, about eight miles upriver from its mouth and directly across the river from the better-known Newcastle upon Tyne, in the northeastern region of England about fifty miles from the Scottish border. The deep and narrow river valley around Newcastle has made bridge building at this location challenging since Roman times, and all early river crossings required the traveler to descend steep roads into the valley and ascend on the other side—until Robert Stephenson's High Level Bridge was completed in 1849. Since the spans needed at Newcastle were not nearly as great as those needed for his contemporaneous Britannia Bridge across the Menai Strait, Stephenson chose a tied arch, or bowstring girder, rather than a tubular girder, for the Tyne crossing's superstructure. The foundation problems encountered beneath the Tyne were more difficult than those at the Menai Strait, however, and numerous timber piles had to be driven through sand, gravel, and clay before the piers could be built. The design and construction were done well, and Stephenson's bridge still carries road and rail traffic across the Tyne.

In the shadow of the High Level Bridge is the Newcastle Swing Bridge, designed by John F. Ure and completed in 1876. A river-level crossing, the 281-foot central span of this bridge pivots in the center to provide wide openings for shipping, thus allowing upriver access. When built, the Swing Bridge was the largest of its type in Great Britain, and it remains a vital traffic link across the Tyne. Just upriver from the Swing Bridge is the Tyne Bridge, a 531-foot-span steel arch that when completed in 1928 was also the longest of its type in Britain. It is this modern high-level crossing that towers over Stephenson's High Level Bridge and the Swing Bridge in the photo accompanying the announcement of the design competition. All three bridges, which are closer to one another than the distances they span, had to be respected by any proposed new crossing of the Tyne.

Details of the 1996 bridge-design competition were provided in a strikingly designed folder, the cover of which reproduced both a photograph looking upstream at the three historic bridges and the ad's eye-catching typography. Inside, there was elaboration on the competition's purpose, program, and procedures. The footbridge, the brochure revealed, was to provide "the essential pedestrian link" between "two major urban regeneration projects." On the Newcastle side, "redevelopment is well underway, with leisure and tourist activities interspersed with high quality commercial buildings and housing." Across the river, "the focus is on cultural and leisure uses," with a former grain warehouse on the riverbank "set to become the largest contemporary visual art gallery outside London." Such was the setting the new footbridge would have to be in keeping with and, moreover, complement.

Potential design competitors were informed that the footbridge "will provide a dramatic new vista of the existing crossings upstream and the sweep of the Tyne as it curves to the south downstream." In addition, the appearance from the other bridges naturally had to be considered in the design. Beyond the aesthetic considerations, however, there were some practical matters to which designers had to attend. The footbridge, which would be located at the convergence of two planned bicycle routes, would have to be accessible to cyclists and accommodate as many as 1.5 million pedestrian and cycle crossings each year. Finally, the bridge would need to incorporate an opening mechanism to allow ships to pass.

The brochure made it clear that multidisciplinary teams would likely be necessary to produce a design "of sufficiently high technical and aesthetic merit." Teams could be led by a civil engineer, a structural engineer, or an architect, "but any contracts or agreements entered into as a result of the competition will be based on civil engineering model documents." To be considered for short-listing, teams were expected to submit the names of all member firms and how they fit into the team's structure; details of qualifications and experience, especially with regard to bridge design and construction; and examples of previous bridge projects, especially opening designs, along with lists of clients. Entries at that stage were restricted to six pages of text on letter-size paper, plus drawings and photographs. Among the criteria for being short-listed were "evidence that the team has understood the unique nature and qualities of the site" and "experience of compatible team working."

Approximately six short-listed teams were expected to be invited to submit a design, which was to be due about six weeks after they were provided further information in the form of a technical brief that would spell out such details as the opening and closing sequences for the bridge's operation, each of which was expected to be completed within two minutes. Each short-listed team would receive two thousand pounds upon receipt of a satisfactory submission of a design, and the winning team would receive an additional three thousand. The winning design would be submitted to the Millennium Commission, a British organization for funding such projects. In other words, a design team was guaranteed no more than two thousand pounds for its efforts, and even the winner might see no more than five thousand, for there was no guarantee of a contract to execute the design.

Though the details may vary, the process for generating ideas for a new bridge across the Tyne is typical of design competitions, which have been held since ancient times. In the fifth century B.C., for example, the Athenian Senate invited architects to submit designs for a war memorial on the Acropolis. The young United States of America held design competitions in 1792 for both the White House and the U.S. Capitol, and in 1802 New York City did so for a new city hall and court-house. In England, a famous design competition for a building to house the Great Exhibition of 1851 drew almost 250 entries, none of which was judged to be adequate. Ultimately the exhibition was housed in Joseph Paxton's late entry of an iron-and-glass structure that came to be known as the Crystal Palace. On continental Europe, in the early twentieth century, the Swiss developed rules for engineering-design competitions that have led to some of the most innovative and attractive concrete bridges in the world. In Australia, on the other hand, the architectural-design competition for a new opera house in Sydney resulted in a striking structure that was recognized worldwide but that, because of its construction problems and cost overruns, effectively ended the use of open design competitions in that country.

Discussions of the pros and cons of design competitions have taken place perhaps as long as there have been competitions. Frank Lloyd Wright believed they led to mediocrity, "an average upon an average by averages in behalf of the average," referring to the compromises made in everything from choosing juries to choosing winning designs. Michael Graves, on the other hand, feels that, although "it would be nice to think one could compete with oneself and that the projects we would do would be just as good if there were no other competitors," design competitions do raise the level of competition and therefore of design. Architects and engineers have long agreed, however, that competitions can be clever, even if unintended, ways to get a broad range of professional services and opinion for a pittance. The late-nineteenth-century civil and mechanical engineer J. W. C. Haldane wrote that "there is no class of work in which engineers, as well as architects, have been so scandalously abused, as that known by the term 'competitions,' " and he devoted a whole chapter in his 1890 book, *Civil and Mechanical Engineering, Popularly and Socially Considered,* to describing and ridiculing the process.

A competition is, of course, not the only way to acquire an excellent design for a new bridge or structure. Larger political entities can support their own design and engineering departments. Many of the bridges built around New York City in the first half of the twentieth century were designed by the city's and the Port Authority's own engineers, who at various times included such outstanding bridge designers as Gustav Lindenthal and Othmar Ammann. On the West Coast, in the 1930s, the state highway departments of Oregon and California had in leadership roles such bridge engineers as Conde McCullough and Charles Purcell, who oversaw the design and construction of the San Francisco–Oakland Bay Bridge, then the largest bridge project in the country. The magnitude of this California bridge project was so great that consulting engineers were necessarily used extensively, but in the final analysis the bridge was but another state-highway project.

Smaller government entities, when not constrained by forbidding procurement procedures, and private organizations can, of course, go directly to an architect or engineer whose work they admire and respect to ask for a design. This does not give much opportunity to young, beginning professionals looking for their first commissions, however, and the design competition does give them the opportunity to show off their talents and possibly capture the ultimate prize of a commission. This was the route that Santiago Calatrava followed for years before getting his first opportunities to have his imaginative paper and model designs realized as full-scale structures.

In the Tyne competition, the six short-listed teams and their designs were described in an illustrated leaflet that was distributed to the public. Comments were invited, and a tear-off sheet was provided for them. The teams and their designs, described here verbatim from the leaflet, were as follows:

Gifford and Partners/Chris Wilkinson Architects: An innovative opening operation combined with a simple design so that whether open or closed, the bridge presents an image of strength and grace.

Napper Partnership/Ove Arup and Partners: Two bridges combined—a rotating bridge on the south side joins a fixed pier supported bridge from the north quay.

WS Atkins/Nicholas Grimshaw & Partners: A slender, cable-stayed deck, pivoting about the location of the support masts, to produce a sail-like structure.

Robert Benaim Associates/Lifschutz Davidson: A two level structure with quay level crossing to provide a pedestrian link between Newcastle and Gateshead town centres, with the possibility of a rapid transit facility on the higher level.

Ove Arup and Partners/Sir Norman Foster and Partners: A flexible deck incorporated into the traditional arch form gives shipping clearance. Twin steel arches support a deck which is pulled upwards to create a third arch when the bridge is opened.

Parkman Ltd./Carlos Casado: A single span link across the river with towers housing the bridge operating system rising from each quayside.

The Gifford/Wilkinson design turned out to be "the most popular with local people," and it was the unanimous choice of the judges, who announced their decision in February 1997. A larger folder then presented the winning design as the Millennium Bridge, quoting Professor Tony Ridley, immediate past president of the Institution of Civil Engineers and chairman of the judging panel:

> A contemporary design which complements the existing Tyne bridges in a way which is both refreshing and new.
>
> Its main arch echoes that of the Tyne Bridge, but it is not a pastiche, it is entirely modern and a fitting tribute to the peak of engineering excellence we are reaching as we near the end of the Millennium.

Whereas the Gateshead Metropolitan Borough Council sponsored the competition, the winning design was supported also by the Newcastle City Council, the Port of Tyne Authority, English Partnerships, and the Tyne and Wear Development Corporation. They were lending their names to endorse the £7.5 million project in the hopes that it would be funded by the Millennium Commission. Although the competition documents had placed an upper limit of fifteen million pounds on the final cost, they had also made it clear that "value for money" was a criterion and that it was "not essential to design a structure costing the max-

imum allowable budget if an equally attractive and appropriate structure can be designed and built for a reduced figure." Nevertheless, on
the announcement of the winning design, at least one local newspaper
report noted that some people had been questioning even the £7.5 million figure. As the paper pointed out, however, the bridge could be had
for half the price of a soccer superstar. This fact was quoted on the
leaflet promoting the chosen concept. Inside the leaflet, which showed
the bridge in various stages of opening, the design was further elabo-
rated on:

> It is created from a pair of steel arches—one forming the deck, the
> other supporting it. Thin suspension rods link the two. The 600
> tonne steel bridge operates like the giant lid of a closed eye slowly
> opening-turning on a pivot on either side of the river bank to form a
> gateway arch which ships sail under.

The competition documents asked for a bridge design "as innovative
as some of its neighbors were when they were built," and the winning
entry appeared to be that. The striking footbridge design resulting from
the competition is truly innovative, and the process for eliciting it
appears to have worked well. By July, a final proposal was being prepared for submission to the Millennium Commission. In the meantime, postcards had been printed showing computer-generated images
of the winning bridge design from various angles and in various stages
of operation. Whether the bridge was ever to be built, however,
depended on whether the funds were to be forthcoming, from the Millennium Commission or elsewhere. Many a striking design and competition winner has remained on the drawing board, computer screen, or
postcard for lack of an ultimate sponsor.

The Gateshead bridge, which ultimately cost twenty-two million
pounds, was funded by the Millennium Commission, and its realization proved to be as much of a mechanical- as a structural-engineering
challenge. In order to rotate the entire 850-tonne bridge on its hinges to
allow ships to pass, a massive hydraulic-ram system had to be designed
and controlled with great precision so that the two sides operated in
unison and raised the bridge and let it down gently enough that it did
not overly oscillate about its rest positions. An entire sequence of open-

Gateshead Millennium Bridge

ing or closing takes about four minutes, during which it moves as slowly as one millimeter per second and as fast as eighteen millimeters per second.

The "blinking-eye bridge," as it has come to be known by the English, is the centerpiece of the Gateshead-Newcastle bankside development area, attracting locals and visitors alike to its striking architecture and spaces. The supporting arch of the bridge is visible over the rooftops on the approach to the area, and its lightness in weight and color is striking. Pedestrians flock to the bridge, stroll across it, often stopping midway to look upriver at the historic bridges that it complements so well. The pedestrian and bicycle bridge has a final unusual feature, a by-product of its curving walk- and cycleway. When this bridge deck is raised, litter that had accumulated on it falls toward the gutter-like edges of the paths and moves along the arc until it reaches traps built into each end. The bridge, like a blinking eye, thus cleans itself every time it is operated.

Proposed drawbridge at Poole, with open leaves mimicking sails

The team that produced the Gateshead Millennium Bridge subsequently entered another competition, this time succeeding with a strikingly original bascule-bridge design for the large natural harbor at Poole, which is a renowned center for sailing located on the southern coast of England. Wilkinson Eyre Architects, as the firm is now known, in collaboration with Gifford & Partners and with the mechanical engineers Bennett & Associates, who were responsible for the all-important opening mechanism of the Gateshead structure, have come up with a bridge for Poole harbor with leaves that are triangular. In the raised position, the 120-foot-long movable spans resemble tall sails, complete with even taller masts, which sit beside the roadway when the bridge is in the closed position. The crossing was expected to be completed in 2006 and should be another eye-opening example of how bridges can be more than functional.

Millennium Legacies

In the late 1990s, while most of the world was anticipating the failure of computers with the dawning of the year 2000, the United Kingdom was building notable structures, like the "blinking-eye bridge," designed to celebrate the millennium. The Y2K problem appeared to be, in the end, just so many system administrators crying wolf, but the British millennium projects, both the successes and the failures, remain as legacies of a noble effort.

A British Millennium Commission was set up in 1993 and was supported with funds deriving from the national lottery. As a source of "good causes" sharing in the proceeds of the lottery, the commission was established to encourage and provide financial assistance for the realization of appropriate projects. Project types suggested for funding ranged from the arts to the sciences, from local village halls to national domes, from passing exhibits to lasting bridges.

Whatever the nature of a proposed millennium project, it had to be distinguished. In the case of a bridge proposal, such as the one submitted by the Gateshead Metropolitan Borough Council, it was generally expected to be the result of a design competition, and collaborations between architects and engineers resulted in some unique structures. Although the cost of a proposed bridge was considered in the judging of proposals, there was also a value placed on what an imaginative design might contribute to the environment in which it would be constructed.

London was a natural location for millennium projects meant to gain international attention as symbols of Britain's technological prowess, but political, economic, functional, and temporal setbacks often brought the wrong kind of attention. The Millennium Dome, which evolved into a gargantuan version of a big top over a multi-ring circus, was conceived of as enclosing a millennium exhibition that would rival London's Great Exhibition of 1851, which was housed in the Crystal Palace, and the Festival of Britain that was held one hundred years later. The Millennium Dome, originally announced in 1994, was canceled shortly thereafter in the face of political opposition but then revived in 1996. In 1997, under Prime Minister Tony Blair's government, the entire project was put under the newly formed New Millennium Experience Company, which was wholly owned by the government because no private investors would accept the risks associated with the ambitious project, and construction was finally begun. Designed by the Richard Rogers Partnership, the dome was to be the centerpiece of the United Kingdom's celebrations, and it was perhaps the most widely reported on of all the millennium projects. Located on the prime meridian on the Greenwich peninsula, the space under the dome was at one point in the early development of the plan conceived to be sectioned off into twelve time zones, in keeping with the significance of its location. As built, the dome was divided into only a half dozen or so "zones" with exhibits on such themes as the human body and spiritual experience.

The Millennium Dome, which has a Teflon-coated roof and is the largest fabric-covered structure in the world, was conceived by the structural-engineering firm of Buro Happold, a worldwide consulting-engineering practice headquartered in the United Kingdom, in Bath. The dome has a diameter of 320 meters (a circumference of a full kilometer) and covers eighty thousand square meters of exhibition, entertainment, and eating space at a cost per unit area said to be less than that of the most economical retail space. Using the customary comparative method employed in describing such large structures, promotional material declared that the entire Eiffel Tower could be laid on its side under the Millennium Dome. The shallow (fifty-meter-high) dome is suspended from 2,600 steel cables attached to a circle of twelve steel masts (each one hundred meters high). As temporary as the masts may appear, they are a permanent part of the permanent design of the structure, which looks not unlike an enormous scalloped shell or a gigantic hubcap suspended from a ring of construction cranes. The dome has also been described as looking like a "partly flattened mushroom punctured by a circle of 12 pins."

Needless to say, as remarkable a structural-engineering accomplishment as it is, the Millennium Dome has had its detractors, and not only because of how it looks. The structure of the dome may successfully keep the sun and rain out, but its contents and operation were generally considered overall to have been disappointments. Guests at the gala New Year's Eve opening, which was attended by the queen and the prime minister, a staunch supporter of the dome, had to wait in long lines to gain admission. The following day, the exhibition was open to the general public, which was supposed to flock to its exhibits, events, and eateries and help provide an infusion of cash to support it financially. A good many people came at first, but the expected crowds were not sustained, and the government had to continue to infuse money into the embarrassing project, which in the end could be justified only because "the reputation of the U.K." was at stake, according to a report by the kingdom's comptroller and auditor general. The total final cost was of the order of one billion pounds.

Though the dome structure itself was to be a permanent fixture, its contents were meant to last only for the calendar year 2000. By the end of February 2001, they were on the public auction block. At first it appeared that the dome itself would be sold to a Japanese firm, but then it was announced that it would be sold to a developer with ties to the Labour Party for £125 million and would be turned into a high-tech business park. Controversy surrounding the arrangement left the future use of the dome uncertain. Finally, in early 2004, it was decided that the dome would be given to American billionaire Philip Amschutz, who was to turn it into a twenty-thousand-seat entertainment venue.

Not all millennium projects were prompted by or supported by the Millennium Commission, but it did inspire others to join the celebration. *The Times* of London, in cooperation with the Architecture Foundation, sponsored a design competition for a landmark for the millennium. Though in the end no winner was chosen, the competition spurred the architects David Marks and Julia Barfield, who now head up the husband-and-wife firm of Marks Barfield Architects, to think about something in the tradition of the Crystal Palace and the Eiffel Tower. Their concept, which was publicized by the *London Evening Standard,* was to erect the largest observation wheel in the world in Jubilee Gardens, on the south bank of the River Thames, between Westminster and Hungerford bridges and just downriver from the Houses of Parliament. When riders reached the top of the Millennium Wheel, as the planned structure came to be called, they would look down on everything else in central London and, on a clear day, would be able to see as far as Heathrow Airport and Windsor Castle, twenty-five miles away.

Although the concept won wide public support, it was some years before planning permission was forthcoming to erect such an enormous structure in the heart of London. At first conceived to be 151 meters in diameter, making it about twice as large as George Ferris's great wheel that towered over the 1893 World's Columbian Exposition in Chicago, the London descendant was finally built with a diameter of 135 meters. (Three years after it opened, the London wheel was still by far the largest-diameter one in the world, bettering the previous record holder by more than half, but larger wheels were already under development in Moscow, Shanghai, and Singapore.)

It was more than sheer size that made the London structure more difficult to finance, erect, and operate than the pioneering wheels of a century earlier. Since it was not supported by the Millennium Commission, the wheel had to have independent financing. In the end, that came from British Airways, which became the principal backer of the wheel because of its potential as a tourist attraction. The structure came to be known officially as the British Airways London Eye, but informally it continued to be called the Millennium Wheel. Whatever it was called, the wheel was said to represent "the turning of the century and the regeneration of time."

At least in part because of the confined space in which it was to be located, the Millennium Wheel was from the beginning intended to be cantilevered out over the Thames. This required a unique design, for which Marks Barfield involved structural engineer Jane Wernick. Her conceptual design resulted in a structure with a main moving part that is not unlike a gigantic bicycle wheel, with its rim made up of a steel space truss formed into a huge circle. The rim is connected to the hub by steel cables, which act like spokes. In the detailed design, which was carried out by the Dutch firm Hollandia, the hub fits over a 2.1-meter-diameter, twenty-five-meter-long spindle fabricated from steel castings, and the spindle is supported by a hinged A-frame that angles out from the riverbank. To maintain its equilibrium, the A-frame is tied back with steel cables.

The wheel has been promoted by British Airways as a truly European project, with the main structure made in Holland of British steel, the hub and spindle cast in the Czech Republic, the bearings made in Germany, the cables in Italy, and the observation capsules in France. The total weight of the polyglot structure is 1,900 metric tons.

Large Ferris wheels have traditionally been assembled in the vertical position in which they will operate. Using temporary scaffolding on a symmetric frame, the rim is typically erected in place, built around the hub by being hung from the bottom up, thus working with—more than fighting against—gravity. Such a procedure was not desirable in the unique London location, where the incomplete structure would have loomed high over public areas, presenting a danger to people below and to workers on the high steel. So, with the help of an enormous floating crane, the wheel was fully assembled in a horizontal position on temporary supports in the river. (This could be done without interfering with boat traffic because at this location the river is twice as wide as the wheel's diameter and the shipping lane is closer to the opposite bank.)

Once fully assembled, the wheel was to be pulled into a vertical position by hydraulic jacks inching steel cables over temporary supports known as shear legs, for which crane jibs were used. The much-anticipated lift, which is believed to be the largest ever of any object from a horizontal to a vertical position, did not go smoothly at first, and problems with temporary cables created a six-week delay. Once that was resolved, it took sixteen hours to raise the wheel to sixty-five degrees off the horizontal, the desired angle of the A-frame. Setting the wheel in the vertical position consumed almost five days, during which time restraint towers that house the drive systems for the wheel also were erected. The initial glitch in the raising of the wheel was a temporary embarrassment, largely forgotten when the monumental task was complete.

Once in place, the wheel was ready to have its passenger capsules—its source of income—attached. There was precedent for such an enterprise being fully self-supporting financially: The world's first Ferris wheel, which could carry a total of 2,160 passengers in its thirty-six trolley-sized cars, made money in its short run of nineteen weeks before the 1893 world's fair closed. The Millennium Wheel, even though significantly larger in diameter than its Chicago ancestor, has only thirty-two capsules holding up to twenty-five passengers each and thus can carry at most eight hundred riders at a time. (The diameter of the capsules was limited to four meters so that they could be transported on French roads between the Alps and the coast in an unescorted "*convoi exceptionnel.*") Still, after only nine months of operation, the London wheel had carried a total of three million riders, twice the number that rode the original Ferris wheel.

The Millennium Wheel achieved this success in part by its year-round operation in a busy part of a busy city and in part by its loading scheme. Ferris's wheel had to be stopped six times to exchange its riders with a completely fresh load. The London wheel does not stop to reload its cars; rather, the passengers exiting and entering the capsules (traveling at only about 0.25 meters per second, which is much slower than normal walking speed) do so on the move, the wheel being stopped only to accommodate the disabled. Also, in contrast to the twenty minutes of intermittent motion above Chicago's Midway Plaisance, the continuous loading of passengers allows riders on the Millennium Wheel to have a thirty-minute experience (one revolution) of smooth, uninterrupted movement. Though continuous boarding is not a new concept—being common with mountain chairlifts and escalators—the utilization of the scheme in London not only makes the operation of the wheel less disruptive mechanically but also makes for a more pleasant and jerk-free ride for the passengers already on board.

When initially outfitted with its observation capsules, which are designed to maintain a horizontal floor for passengers to stand on, the wheel naturally had to be tested for safety before it could be opened to the public. Tests revealed that the clutches on the capsules were faulty, and they all had to be replaced. The work further delayed the opening of what had come to be called the "wheel of misfortune." Rather than carry its first passengers as planned for the eve-of-the-millennium celebration, the wheel turned empty, illuminated by a laser light show. The Eye finally opened two months past schedule, but it has since operated more or less smoothly, being closed down only infrequently for minor computer and mechanical problems. British Airways expected to operate it for at least five years, but many Londoners hoped that what was conceived as a temporary structure would, like the Eiffel Tower, become permanent.

When my wife and I visited the Eye on an overcast morning in the summer of 2003, we found that it was undergoing painting—a sure indication that it had indeed been granted an extended run. People flocking to the attraction in numbers beyond all estimates make the area an exciting place to visit and board a "flight," as a circuit on the airline-sponsored wheel is marketed. Our capsule with a full complement of passengers was not at all crowded and was extremely comfortable throughout the flight. There was plenty of space to move about, enabling us to look in all directions at a sprawling London beneath us. Well before we reached the apex of the journey, we experienced the pleasant sociological phenomenon of people confined to a closed space becoming friendly and cooperative, offering to take pictures of one another's group and posing them against the best background. The only negative experience occurred as passengers waiting to board watched crews search on the run each just-emptied capsule for bombs and other malicious things a terrorist might have left behind. However, as soon as the next group of passengers was loaded, such unpleasant thoughts were soon forgotten in the slow, silent motion of the wheel.

The smoothness of the wheel's operation belies its difficult beginnings, however, and a surprising literary embarrassment. After almost one million people had ridden the Eye, a passenger finally noticed and reported that some lines of poetry were incorrectly quoted on a plaque beside the wheel. The lines are from William Wordsworth's sonnet "Composed upon Westminster Bridge," and they are apt not only because the wheel is located near the bridge but also because of what they say:

> *The river glideth at his own sweet will:*
> *Dear God! the very houses seem asleep.*

The plaque beside the wheel had compressed the two lines into a single unintelligible one: "The river that glideth at his own sweet asleep." After the error was pointed out, the plaque was quickly corrected, but no one could explain how the mistake had been made in the first place or why it had not been caught sooner. Such questions are often asked also of engineering designs that exhibit behavior that takes everyone by surprise, as it did in another much-anticipated London millennium project.

The London Millennium Bridge—like similarly named bridges elsewhere in the British Isles—is strictly for pedestrians. It was proposed to connect the staid architecture of the financial district of the City of London, at a location near St. Paul's Cathedral, with the new Tate Modern art museum, which is across the Thames at Bankside, in reclaimed space within a landmark powerhouse. (The Tate Modern is itself a millennium project.)

An international competition for a bridge design was sponsored by

London Millennium Bridge, leading to St. Paul's

the *Financial Times* and run by the Royal Institute of British Architects. The London-based team comprising Lord Norman Foster's architectural firm, Foster & Partners, the engineering firm Ove Arup & Partners, and the sculptor Sir Anthony Caro claimed the prize. Their design is a 320-meter-long suspension bridge with a 144-meter main span and low Y-shaped towers. The low towers support a correspondingly low suspension system consisting of twelve-centimeter-diameter steel cables spread horizontally in groups of four on each side of the four-meter-wide pedestrian walkway. Because the profile of the bridge is so shallow, with a maximum cable sag of only 2.3 meters, the cables are necessarily highly tensioned, which affects the natural frequency of the structure.

While it is still a set of drawings on paper or a computer screen, such a complex structure is routinely checked by engineers for strength, stiffness, and stability through the use of computer models. The natural frequencies of the structure are determined, and its response to various loading conditions is checked to be sure that no sympathetic modes of vibration can be expected to be excited and amplified beyond safe limits. In the course of designing the London Millennium Bridge, Arup engineers assumed that people using the bridge would exert vertical forces on it but that they would not be in step. The possibility of pedestrians exerting significant horizontal forces in unison was not even considered. The assumptions made by the engineers were customary and hardly worth a second thought. In fact, the form of the bridge seemed much more innovative than its function, and so its artistic and architectural elements may have gotten more attention than its engineering.

The materials of the Millennium Bridge are aluminum and stainless steel, and it has been frequently described as looking "high-tech." Its location and the absence of vehicle traffic were expected to draw four million people annually to use it. Because the Millennium Bridge was the first new river crossing built in the area since Tower Bridge was completed in 1894—and the first pedestrian-only bridge—opening day was a much-anticipated event. When the day arrived, on June 10, 2000, almost one hundred thousand people eventually showed up to admire the bridge and walk across it. To everyone's surprise, the lightweight bridge deck swayed noticeably side to side. The bridge was closed within three days, after being used by as many as a quarter of a million pedestrians, and it remained closed while engineers sought to understand the motion and work out ways to check it.

A bridge moving under the footfalls of people is not a new phenomenon. Many a hiker has experienced how easy it is to set into motion a light, narrow bridge suspended over a mountain ravine. Nineteenth-century suspension bridges were known to have been brought down by the cadenced march of soldiers, and to this day some spans display a sign warning soldiers to break step when crossing. London's Albert Bridge bears the terse notice "All Ranks Break Step." As recently as September 1999, a new pedestrian bridge in Paris exhibited excessive sideways motion under the feet of walkers. The Pont de Solferino, a 106-meter-long narrow steel arch spanning the Seine near the Louvre, was closed shortly after it opened.

Clearly, it was the footsteps of pedestrians that were exciting the Millennium and other bridges, but why they were doing so was not immediately obvious. The expectation was that large crowds of people would subject the bridge to random vertical forces, favoring no particular frequency and hence exerting no strong forcing function on any particular natural frequency of the bridge structure. But people on a crowded narrow bridge cannot walk at random speeds; they must fall in step with the general flow.

Investigating the misbehavior of the Millennium Bridge fell to Arup and the consultants the firm engaged. The resources of the Institute of Sound and Vibration Research at the University of Southampton were called upon, and a physical model of a sideways-moving walkway was constructed at Imperial College, London. The effects of a person walking over it helped engineers understand the nature of the forces of interaction. A mechanical shaking device was installed on the actual bridge to excite and verify the nature of the vibration modes. At one point, a corps of Arup employees was dispatched to the Millennium Bridge to conduct tests upon it. They grouped together and moved in unison, presenting a sight reminiscent of the load test of a section of the galleries of the Crystal Palace conducted 150 years earlier in London's Hyde Park. Over the Thames, it was found that it took only five hundred to six hundred walkers to set off the wobble observed on opening day, when there had been as many as two thousand people on the structure at one time.

As it turns out, a normal walking pace is about two strides per second. Thus, the two footfalls per second exert a vertical force at a frequency of two cycles per second on the surface over which pedestrians walk. This is the force that was taken into account in the design of the bridge. There is also a backward horizontal component of force exerted at the same frequency by a walker; otherwise there could be no forward movement. Another component of force, however, is more subtle: a sideways horizontal push to counter the natural side-to-side motion of the human body walking. This sideways force alternates left and right, so in each direction it is exerted at only half the frequency of the other forces, and normally it is relatively small. However, the frequency of this one-cycle-per-second force matched a natural frequency of the Millennium Bridge and set it into vibration in an unexpected S-shaped mode in the plane of the walkway.

The principal finding of the various studies was that the Millennium Bridge exhibited a pedestrian-structure interaction, or "crowd-induced dynamic pedestrian loading." When the bridge was being used by walkers moving in random ways, it was possible that by chance a sufficient number of them were in step and so caused the walkway to move sideways, if only slightly. However, since the motion takes place at a relatively high frequency, and since humans tend to be sensitive to higher-frequency motion, the slight movement was amplified in their minds and bodies. Reacting to the motion sensed, the people on the bridge at the time began to move side-to-side in synchrony with the bridge deck, thus amplifying the motion much as a parent pushing a child in a swing increases the arc with each cycle. In time, the amplitude of the movement of the bridge became visibly more and more noticeable and psychologically more disconcerting to those on its walkway. On the bridge's opening day, there was also a high crosswind that caused walkers to brace themselves against it and so exert in opposition to the wind a stronger-than-normal horizontal-force component on the bridge, thus amplifying the latent effect. The resultant wobble of the walkway caused hundreds of people on the bridge to grasp its handrails to steady themselves.

What happened on opening day could naturally happen again, and so the bridge structure or the forces to which it could be subjected had to be altered. Limiting the number of people on the bridge at any one time was ruled out, as was altering their walking patterns by installing "street furniture" and other obstructions along the walkway. These solutions were considered not in keeping with the intended free use of the bridge.

If the load exerted on the bridge was not to be controlled, then the bridge structure itself had to be altered. One way of doing this was to stiffen the bridge structurally. This would involve major construction work, would be costly and time-consuming, and would alter the appearance of what was, after all, a piece of architectural sculpture as well as a utilitarian structure.

The third way to prevent unacceptable motion of the bridge was to install mechanical dampers, which would dissipate energy and limit movement. Two kinds were considered. The first were viscous dampers, which act as shock absorbers, like those installed on the Pont de Normandie. The second alternative was to use tuned mass dampers, in which a large mass is connected to the structure by stiff springs. Tuned to the desired frequency, the mass and spring can absorb much of the energy that would otherwise go into exciting the bridge.

In the end, it was decided that both kinds of dampers would be installed on the Millennium Bridge in the most unobtrusive way possible, and the deck structure was also to be stiffened against lateral deflection. Each of these solutions would necessarily alter the appearance of the bridge, but the alterations would be largely invisible to those walking across it. Everything, however, would be visible from underneath by the people on boats that cruise up and down the Thames. The strong rectangular pattern of the underside of the bridge deck would be interrupted by the cross-bracing used to stiffen the deck, tuned mass dampers would interrupt the slenderness of the cross members, and the struts of the viscous dampers at the piers would also be visible. The final design for the fix employed a total of ninety-one dampers of various kinds, but the bridge was not stiffened structurally, so as not to alter its appearance any more than necessary. The work was completed about two years after the bridge was initially opened, and though a small amount of motion could still be detected by those walking across it, the damping system has kept that to a minimum.

The problems experienced by London millennium projects were embarrassing to British planners and engineers. However, each of them was clearly pushing the envelope of experience and testing the limits of human ingenuity in both engineering art and engineering science. It is the intent of visionary designers to work on the edge of technology, and it is the nature of their creations to awe others, engineers and nonengineers alike. Not every concept proves to be as exciting when realized; not every design works as smoothly as conceived. Not every wheel turns as freely, nor does every bridge stand as steady as it should. But if there never were an embarrassment, could any engineers honestly say that they had been trying as hard as they might to celebrate the turning of a millennium or to lift the spirits of the world?

Broken Bridges

Shortly after the 1994 Los Angeles earthquake struck, a veteran engineer called me to talk about the broken bridges, pictures of which were ubiquitous in television and newspaper reports of the disaster. In the fallen spans, which had come to epitomize the destructive and disruptive force of the quake in auto-centric southern California, he saw both classic and confusing examples of the failure modes that so dominate engineering-design considerations. Since the objective of engineering is to identify and obviate failure modes in the design of structures such as bridges, when failures arise unexpectedly they are naturally the object of close scrutiny. What caused the classic and unfamiliar modes of failure? the engineer wondered. Such questions, which occur to engineers and nonengineers alike in the wake of failures, not only help us to understand the state of the art of engineering but also help to illustrate its nature and limitations. The questions, however, are often articulated more easily than are definitive answers.

Among the things the old-time engineer wondered was whether sufficient information about the failures was going to be collected before the debris was cleared, and how effectively such information would be disseminated to the wider engineering community, young and old alike. Because the design and construction of large civil structures, like the major highway bridges that failed around Los Angeles, take place in a context of theory and practice that is evolving constantly and that is never tested completely, any failure presents a rare opportunity for a fortuitous experiment to test hypotheses on which bridge designs are necessarily based. Unfortunately, just as economic and physical considerations argue against full-scale tests of large structures prior to their construction, the same considerations severely limit the time frame in which information about the failures can be collected.

The engineering ideal would be to leave the broken bridges untouched so that engineers could pore over old calculations and carry out new ones while they designed and performed tests on computer models, scale models, and even full-scale models provided by the adjacent but unfailed bridges. As calculating and testing progressed, naturally revealing new insights and speculations about what exactly happened during the earthquake, the failure site could be revisited to check developing theories against reality.

A model approach is followed in the wake of airplane failures, in which case the National Transportation Safety Board (NTSB) has the authority to cordon off a crash site so that every piece of debris can be collected, arranged, and preserved in a nearby hangar until the investigation is completed. Even then, recognizing that no matter how careful an investigation and no matter how incontrovertible an explanation for a crash may seem, the NTSB declares only a "most probable cause" for a crash.

Although such a failure-analysis procedure can be imagined for fallen highway bridges, the practical constraints against it are overwhelming. Who, for example, could argue effectively that the broken bridges should remain as they fell on the freeways around Los Angeles (because the parts are too massive to move intact) until engineers are satisfied that they have collected all of the information they want? Pressures dictate that the debris be broken up and removed as quickly as possible so that the heavily traveled highways can be reopened with a minimum of delay.

Civil-engineering structures differ from aeronautical creations in that the former can never be so readily tested under controlled conditions as the latter. Because aircraft are maneuverable by design, it is economical and feasible, not to mention prudent, for new designs to be subjected to extensive test-flights. Furthermore, because airplanes are built in numbers and operated in fleets, having one out of commission (because of an accident or for maintenance or repair or testing) presents no significant interruption in service. A major bridge, on the other hand, can practically be fully shaken only with the force of actual wind or a real earthquake, which is of course an uncontrolled and unplanned experiment, and the most ambitious bridges are usually built as unique examples. (Some rare exceptions are the parallel pair of virtually identical suspension bridges, built seventeen years apart, carrying I-295 across the Delaware River, and the twin drawbridges that will replace the Woodrow Wilson Bridge, which carries I-95 across the Potomac.) Computer and scale-model tests are possible and can be performed on bridge designs, of course, but these are always subject to the limitations of being tests of models of the structure rather than of the real structure itself.

Highway-bridge failures have seldom claimed as many lives as an airplane crash, but there have been exceptions. The collapse of the double-deck highway in Oakland, California, during the 1989 Loma Prieta earthquake claimed forty-two lives, and the number of dead and injured during the 1994 Los Angeles earthquake might have been considerably greater had the incident not taken place so early in the morning.

The greatest losses of life in bridge accidents in America have typically arisen not from natural disasters but from faulty or poor design or construction. More than eighty people died when the Ashtabula Bridge in Ohio collapsed in 1876 while a railroad train was crossing it in a snowstorm. Although the cause of the accident is still the subject of discussion, it is generally believed to have been due principally to faulty cast-iron construction of the truss bridge. In 1967, the Point Pleasant Bridge across the Ohio River collapsed suddenly, killing forty-six of the people caught in rush-hour traffic between Gallipolis, Ohio, and Point Pleasant, West Virginia. The failure was traced to corrosion-assisted cracking that had progressed undetected in the eyebar chain links, in a location that was largely uninspectable. Some of the most famous bridge failures, such as that of the Tacoma Narrows, claimed no human lives.

Can bridges be made perfectly safe? In particular, can they be designed to be totally immune to major earthquakes? Although such an ideal state is possible in theory, it is highly unlikely for several reasons, including economical, political, aesthetic, and technical ones. The economics of the situation are such that highways and bridges that could withstand all earthquakes would cost a great deal of money and so would have a severe impediment to getting built in a political climate where there is considerable competition for scarce financial resources. The realities are that bridges are designed for a certain kind and level of earthquake, which seems at the time to be the maximum credible one.

The engineering problem, simply stated, is always to design a good-looking bridge that is both safe and economical. Like all engineering-design problems, this one is fraught with conflicting objectives that have to be reconciled in an imperfectly understood universe of choice. There is no science of bridges, nor can there be, if science means the mastery of what already exists. The very act of designing a bridge involves creating what does not exist, at least in the exact form required for a new location. Engineering design is a self-referential process, and there is no differential equation epitomizing bridgeness to which one need only append appropriate boundary conditions to find a unique or even just a neat mathematical solution. The conceptual design of a bridge, even the most common of highway spans, begins in an engineer's mind, emerging from an ineffable synthesis of bridges that he or she has seen, heard about, read about, or dreamed about. Only when an engineer captures this image on paper or on a computer screen can the design proper begin. It is at this stage that equations can be written, theories applied, and questions of failure modes and prevention asked and answered with some degree of precision and quantification.

Although ground conditions and potential earthquakes constitute only two of the considerations that must go into bridge design, they can dominate discussion at times. When the Golden Gate Bridge was being designed in the early 1930s, for example, considerable differences of opinion between geologists and engineers arose over the question of foundation conditions. Yet, unlike the nearby San Francisco–Oakland Bay Bridge, a section of the upper deck of which collapsed onto its lower roadway during the 1989 earthquake, the Golden Gate Bridge was undamaged in that seismic event, which had its epicenter eighty miles away. Nevertheless, after that earthquake, the State of California mandated that the Golden Gate Bridge be made strong enough to remain functional after an earthquake measuring 8.3 on the Richter scale. Preliminary studies of how the bridge would respond to such a large quake centered closer to San Francisco indicated that the cables might slide off the towers, the side trusses might slam into the towers, and the towers themselves might rock on their foundations. Exactly what might happen would depend, of course, on the exact magnitude and direction of the earthquake, and therein lies one of the most difficult aspects of design for seismic loads.

When the San Francisco–Oakland Bay Bridge was being designed in the early 1930s, its chief engineer, Charles H. Purcell, wrote that "unusual precautions" had been taken to safeguard against earth shocks for what would be the world's longest bridge. What "earth shocks" meant quantitatively, of course, provided the input for the analysis of how the design would respond. According to Purcell, "all elements of the bridge were designed for an acceleration of the supporting material of 10 per cent that of gravity." With the further assumption that the peak acceleration will be parallel to the bridge, for example, it can be determined whether or not a section of the bridge deck will move off its support and fall. According to calculations, that would not happen, but the 1989 San Francisco earthquake evidently produced ground motions greater than those assumed in the 1930s, and a key Bay Area traffic link was broken. After that earthquake, the slender supports of the bridge were retrofitted with tapered concrete jackets to give the structure greater stiffness and thus not magnify the ground motion in future quakes; studies for bringing the entire East Bay portion of the bridge up to acceptable earthquake standards were soon begun.

Engineering assumptions about the magnitude and direction of an earthquake are based largely on the historical record of seismic events and on geological knowledge of fault lines and locations. The Northridge earthquake of 1994 was traced, after the fact, to a previously unknown fault of the kind known as a blind thrust, which tends not to break the surface of the earth even though large amounts of energy are released from slipping at a steep angle to the horizontal. The unique characteristics of such a fault were demonstrated by recordings made by the California Strong Motion Instrumentation Program at various distances from the epicenter. About five miles away, at Tarzana, for example, maximum accelerations reached 1.82 and 1.18 times that of gravity in the horizontal and vertical directions, respectively. Not only are such accelerations more than an order of magnitude greater than those originally assumed in the design of the Bay Bridge, but the strong vertical component showed the earthquake to be of an unusual type, one in which structures like bridges would be shaken considerably not just horizontally but vertically.

An unusually strong vertical motion coming from a previously unknown fault is not the kind of condition that an engineer can readily design against. If highway engineers had in fact proposed taking into account such extreme conditions when the roads and bridges in the area were being designed, they might reasonably have been expected to justify their assumptions. Without a history of strong vertical ground motion and without a knowledge of a blind thrust deep below the ground surface in the area, it is highly unlikely that plans for bridges designed to withstand such incredible-seeming conditions would have been approved and built.

Even after the Northridge earthquake broke key links in the Los Angeles freeway system, there remained some open questions about what caused some of the structural failures. A column failure on the Simi Valley Freeway, for example, revealed that the steel reinforcement deformed into a basketlike arrangement of severely bent rods. Some reports suggested that this failure mode was caused by rocking horizontal motion, which caused some compressed bars to bend outward, thereby pushing off their concrete cover, which in turn so weakened the column that further rods bent outward in the subsequent earthquake motion. It is also possible that the failure resulted primarily from the vertical bouncing of the bridge deck on the column, also causing it to swell outward, again with the rods pushing out the concrete, but more or less simultaneously and in a more symmetrical fashion.

Establishing exactly how the failure mode originated and progressed was further complicated by the fact that the overpass in question crossed the streets below at an angle, thus leading to a design in which some lines of columns were skewed to the roadway. The various combinations of horizontal and vertical ground motion, combined with the relatively complex geometry of the problem, cause the analysis of the structure to proceed in different directions depending on the assumptions made at the outset. With the understandable pressures to repair and reopen the freeways as soon as possible, the physical evidence of the failure modes of the various columns could be retained only in photographic records. Although these help to confirm or refute the various failure scenarios that might be offered, the loss of the failed artifact itself always leaves some room for debate about what exactly happened.

Regardless of the precise sequence of events that causes failures like those of the Simi Valley Freeway columns, it is possible for engineers to retrofit similar structures. What is clear, even from photographs, is that the steel reinforcing bars in the concrete columns deflected outward. Even before the Northridge earthquake, the California Department of Transportation had instituted a program of retrofitting bridge columns with steel jackets capable of containing the outward pressures that improperly reinforced concrete cannot. Bridges that had been so retrofitted appear to have survived the earthquake intact. However, the Simi Valley Freeway, which was built just after the 1971 Los Angeles earthquake, had what turned out to be an insufficient amount of spiral reinforcing to contain the vertical steel. But since it was not known that the structure was directly above the line of a deep thrust fault, retrofitting bridge columns in that location had not been given the highest priority.

At any given time, engineering design and analysis are only as good as the knowledge and assumptions upon which they are based. Whether a bridge is declared safe or unsafe against the next earthquake can be stated with certainty only if the nature of the earthquake itself is known with certainty. Since that kind of knowledge is unlikely in the foreseeable future, engineers must design and analyze new and existing bridges with a degree of judgment that will remain open to debate and scrutiny. Nevertheless, engineering the safest bridges possible is a challenge that has by and large been met, for no damage was detected on any of the 114 bridges that had been retrofitted before the Northridge earthquake. The lessons learned from it and from other seismic events, however limited, enable engineers to design new bridges and other structures better able to withstand future earthquakes.

New and Future Bridges

An era of great bridge building in America reached a peak in the 1930s, when the George Washington, Golden Gate, and San Francisco–Oakland Bay bridges were completed. The first two of these had record-setting main suspension spans of 3,500 and 4,200 feet, respectively, whereas the third, costing seventy-five million Great Depression dollars, was the longest and most expensive bridge project to date. Many other structurally significant but lesser-known spans were also built in those hard times, but the era ended abruptly in 1940 with the collapse of the Tacoma Narrows Bridge. The dramatic failure was recorded on film, which became incorporated into newsreels and embedded in the minds of people everywhere as a symbol of the hubris of engineers.

In addition to the obvious fact that bridge designers could no longer claim complete knowledge of their art, there were other factors that contributed to the subsequent hiatus in long-span bridge building. That interruption, which would last for a decade, was also brought about by World War II and the rationing of crucial materials that accompanied it. Furthermore, the most critically needed fixed links for carrying the country's growing automobile traffic were then in place. However, there were great bridges still to be built.

Although a crossing between the lower and upper peninsulas of Michigan would have alleviated long lines of cars waiting for ferries, the traffic there comprised mainly hunters and vacationers and was thus seasonal and not so pressing a problem to be solved. Another obvious location for a bridge was at the entrance to New York Harbor, between Brooklyn and Staten Island, but the latter borough of New York City was not very highly populated at the time, so that bridge also could wait until the war ended and the Tacoma Narrows collapse could be convincingly explained.

A redesigned Tacoma Narrows Bridge opened in 1950, and the Mackinac Straits Bridge eventually was built, opening in 1957, with special design attention paid to how the deck would behave in the wind. Unlike the original Tacoma Narrows Bridge—and the George Washington Bridge, whose state-of-the-art design it had followed—the Mackinac employed a very deep deck truss and an open-gridded roadway portion to allow wind to pass through the bridge rather than wreak havoc on it. (It was claimed that this feature made the bridge stable in any wind.) The Verrazano-Narrows Bridge across New York Harbor was completed in 1964, with its lower deck in place from the beginning—not because traffic demanded it but as a conservative design measure to stiffen the world-record 4,260-foot central span. Even before the completion of the Verrazano-Narrows, America had all the top-ranking suspension bridges in the world, but it would not retain that distinction for long.

When the 4,626-foot Humber Bridge was completed in England in 1981, America lost its claim to having the world's greatest suspension bridge, at least as measured by length of main span, and the United States is not likely to recover the title in the forseeable future, if ever. (The longest suspension bridge built in America in the last forty years is the 2,390-foot Alfred Zampa Memorial Bridge, completed in 2003 across Carquinez Strait at the northern end of San Francisco Bay.) The country's great rivers, straits, and harbors have by and large all been bridged, and, although arguments could be made for additional spans across, say, the Hudson River or San Francisco Bay, the best locations technically have already been developed. Thus, new spans would face great design challenges, which often translate into prohibitively expensive solutions. Furthermore, the changed regulatory climate might not allow new spans to be built at all. Conditions may change in the future, but the present climate suggests that it is likely to be a distant future.

The future has already arrived elsewhere, however, and the closing years of the twentieth century saw the opening of what were the two largest suspension bridges in the world. In June 1998, Denmark officially inaugurated its Great Belt Bridge, which crosses the 4.2-mile-wide strait between the islands of Fyn and Sjælland, on which Copenhagen is located. Since Fyn was already connected across the narrower Little Belt to the Jutland peninsula, Copenhagen was finally linked to the mainland. (It was only a few years later that the cosmopolitan city was also connected to Sweden, when in 2000 the Oresund Crossing—a ten-mile-long fixed link incorporating a tunnel, an artificial island, and a 3,325-foot-long cable-stayed bridge with a 1,495-foot-long main span—was completed.) The suspended part of the Great Belt Bridge, with a 5,328-foot-long main span and its 8,838-foot total length between anchorages, could have made the bridge the longest in the world on both counts, but those distinctions belong to a Japanese bridge.

The Akashi Kaikyo Bridge, which opened in 1998 near Kobe, is likely to be the longest suspension bridge for some time. Its towers were in place when a 1995 earthquake hit, and it left them a couple of feet farther apart but otherwise undamaged. Since the roadway had not yet been hung from the cables, its components were redesigned to fit properly into the main span, which is 6,529 feet between towers. The 12,828-foot length of the total suspended span means that drivers travel almost 2.5 miles between the 350,000-ton anchorages.

The span across the Akashi Strait is only one part of the massive Honshu-Shikoku Bridge Project, which links the islands of Honshu, on which Kobe is located, and Shikoku by way of the small island of Awaji. The bridge has been designed to withstand an earthquake of magnitude 8.5, and its survival of the 7.5 Kobe earthquake, albeit while still under construction, provides some confirmation of the design's integrity but raises serious questions about the wisdom of building colossal structures in earthquake zones.

Such experience may have severe implications for the design of a fixed link in another seismic zone—across the Messina Strait between the Italian peninsula and Sicily, a project that was discussed at least as early as 1870, when a rail tunnel was proposed. An international competition was held in 1969 to encourage conceptual designs, and more than 150 entries were submitted, about 90 percent of them from Italian engineers and firms. Among the schemes to cross the strait between the presumed sites of the mythological Scylla and Charybdis, those taken most seriously were tunnels and suspension bridges. Tunnels, however, are costly and would take a long time to construct at this location, and so the bridge proposals have continued to attract the most attention. One design calls for a bridge with a 9,500-foot main span, necessitated by the tower foundations being sited on the shores. The Messina Strait bridge would also be three times as long as the longest suspension bridge previously designed to carry railroad traffic, the Tagus River Bridge in Lisbon, Portugal, showing further the uncharted waters into which daring projects enter.

The deck design for the potential record-setting span is of the wing-like box-girder type favored by the British. This is not surprising, since among the structural consultants was William Brown, who for many years was with the firm of Freeman Fox & Partners, which developed the girder design for its Severn and Humber bridges in England and copied it on such other projects as the Bosporus Bridge in Istanbul. (The Akashi Kaikyo Bridge, on the other hand, adopted the American deck-stiffening truss structure characterized by such familiar bridges as the Golden Gate and the Verrazano-Narrows.)

With structures as far beyond experience as the proposed Messina Strait design, unusual features sometimes have to be adopted. For example, with the large torsional deflections (twisting motion) possible in such a long bridge span, railroad tracks must be located along the centerline. The outside vehicle roadways are designed so that traffic approaching the bridge in the Continental mode of driving on the right will be repositioned to proceed across the bridge on the left side, in the British way. The rerouting was proposed so that motor vehicles would move in the same direction as adjacent rail traffic so as to minimize wind effects from passing trains. Vehicles would, of course, have to be rerouted again after they leave the bridge. In any case, whether this or any bridge ever crosses the Messina Strait will depend, not surprisingly, on the political and economic climate in Italy and elsewhere.

A much more ambitious superbridge proposal is for a fixed link across the Strait of Gibraltar, an intercontinental link that is certain to be as politically complex to bring to fruition as the Channel Tunnel. However, the economic and social implications of uniting Africa and Europe across the entrance to the Mediterranean led the United

Nations, through UNESCO, to initiate in the late 1970s studies for such a crossing. An interim report prepared under the auspices of separate economic commissions for Africa and Europe, with the cooperation of the governments of Morocco and Spain, was transmitted to the UN's Economic and Social Council in the mid-1980s, and it summarized the findings of consultants and the status of ongoing studies. International congresses were held to discuss the various aspects of such an unprecedented project, and discussion continues.

As with large engineering projects generally, there are endless possibilities for the location and design of a structure to cross the Strait of Gibraltar, but experience and engineering judgment have narrowed the field to a few realistic choices to be studied in some detail and compared. In the case of a Gibraltar crossing, this task fell to the lead consultant, Freeman Fox & Partners, which identified three technically and geographically sensible locations, ranging from a nine-mile route passing over water as deep as three thousand feet to a twenty-seven-mile route over somewhat shallower depths. In the final analysis, both of these routes were rejected in favor of one that covers a distance of nineteen miles over depths not exceeding 1,150 feet. This still presents enormous challenges well beyond firm engineering precedent for either a tunnel, a floating bridge, or a fixed-support bridge—or a combination of these structures. Since it is recognized that technological advancement in the state of the art of construction proceeds even as studies are carried out, it has been recommended that the final choice of structure not be made until the decision to go ahead with the project is at hand.

Since engineering, like science, progresses through peer review, the report of Freeman Fox was reviewed by an equally prestigious consultant, the Bechtel Corporation, which in turn engaged the world-class bridge-designing firm of T. Y. Lin International to study the feasibility of constructing a bridge or bridges across the strait. Because of the extreme depths of the water and the high volume of shipping at the location, a clear objective was to have as few piers as practicable, which obviously meant having the longest possible bridge spans. This led T. Y. Lin himself to propose a bridge across the nine-mile route that would have only three piers but in water as deep as 1,500 feet and with spans as large as sixteen thousand feet. Although this is well beyond the length

Proposed bridge across Strait of Gibraltar

of the main span of the Akashi Kaikyo Bridge, the generally agreed-on feasibility of the proposal supported the hybrid design put forth by Lin.

The design combines features of suspension and cantilever bridges, with the suspended span being about ten thousand feet long and the cantilever arms reaching out about 3,300 feet each. The design is reminiscent of the 1920s proposal for an ungainly cantilever-suspension bridge that nonetheless enabled Joseph Strauss to promote the creation of a Golden Gate Bridge and Highway District, a crucial step in getting that bridge built. The original ungainly design for the California structure evolved into what many consider still to be the most beautiful bridge in the world. Lin also proposed as an alternative design for Gibraltar a combination suspension and cable-stayed bridge but recognized, as Strauss eventually did, that "the aesthetics of this system can stand some improvement." The fundamental thing in the conceptual design stage is to establish feasibility within, or even reasonably and defensibly beyond, the state of the art.

As an alternative to a bridge, a tunnel beneath the Strait of Gibraltar was proposed as early as 1869, the year the Suez Canal was completed. Large projects have always emboldened engineers to propose still larger projects, and in recent years the completion of the Channel Tunnel has renewed interest in a tunnel scheme between Spain and Morocco. Exploratory borings have been made in Morocco, to depths of one thousand feet. The clay found there is common to the two coastlines, but what lies in a tunnel's path under the strait cannot be known with absolute certainty until the actual tunnel is bored. Likewise, even though the technology of offshore oil platforms may be adopted to set bridge piers in unprecedented depths of water, what surprises might be in store at those depths will remain unknown until construction actually begins. Such uncertainties, not to mention the multibillion-dollar cost estimates, are likely to escalate as the realities of a unique construction project reveal themselves, as they did with the Channel Tunnel, thus keeping a fixed crossing of the Strait of Gibraltar on the drawing board.

T. Y. Lin made his proposals for a bridge across the warm waters of the Mediterranean as a subconsultant, but his involvement with a projected bridge across the frigid, ice-laden waters of the Bering Strait was as a leader. The idea for a fixed link between Alaska and Siberia arose when the Alaskan territory was about to be acquired by the United States in the 1860s. Because Western Union did not believe a transoceanic cable was practical, it dispatched survey and construction crews to find an overland route for a five-thousand-mile telegraph line between the United States and Siberia, which would have required the construction of a bridge over the Bering Strait. The plan was set aside when, in 1866, Cyrus Field did demonstrate the feasibility of a transatlantic cable. Still, the ascendancy of the railroad provided a continuing impetus for pursuing the idea of a bridge between North America and Asia.

A "cosmopolitan railway," which had the object of "compacting and fusing together all the world's continents," was the subject of an 1890 book by William Gilpin, the first territorial governor of Colorado. That same year, on the occasion of the opening of the Forth Bridge in Scotland, an illustration in a souvenir program showed a locomotive labeled "Progress" pulling a passenger coach with a visionary route identified as "Aberdeen to New York, via Tay Bridge, Forth Bridge, Channel Tunnel, and Alaska." A young Joseph Strauss, long before he would become the mastermind and chief engineer of the Golden Gate Bridge, proposed an international railroad bridge across the Bering Strait in his 1892 University of Cincinnati graduation thesis.

The 1887 decision to extend the Russian railway system into Siberia and the further step, in 1891, to begin construction of a Trans-Siberian Railway (finally completed a century later) encouraged the idea of a bridge. On the other side of the strait, the discovery of gold in Alaska made a fixed crossing even more desirable, as did the continuing development of the railroads. Yet fifty years later, the route that is believed once to have been the site of a land bridge by which people, plants, and animals migrated to the North American continent still remained water locked. Enthusiasm for the idea was reawakened during World War II, when the United States and the Soviet Union were allies, but the cold war that subsequently developed cooled people to the prospect of a back-door link around the Iron Curtain.

It was in 1958, the year after *Sputnik* accelerated engineering competition between the superpowers, that bridging the Bering Strait again came to the fore as an ameliorative. Lin, in conjunction with Washington senator Warren Magnuson, who was chair of the Senate Committee on Commerce, suggested a bridge across the Bering Strait "to foster commerce and understanding between the United States and the Soviet Union." (Shortly thereafter, Russian engineers suggested a dam be built across the strait to regulate the flow of water between the Arctic and the Pacific oceans and thus control weather, but there was disagreement as to which way the water should be pumped.) A decade later, Lin organized Intercontinental Peace Bridge, a charitable corporation created "to initiate and perpetuate efforts toward the eventual building of this hemispheric link." As just about every other major bridge proposal has had its tunnel counterpart and vice versa, so the Bering Strait bridge effort has the Interhemispheric Bering Strait Tunnel and Railroad Group, headed by the engineer George Koumal, who in the late 1980s "decided to pursue the concept with a vengeance."

Although the successful completion of the Channel Tunnel renewed interest in massive projects, the cost overruns and tenuous financial condition of the Chunnel have made it more difficult for tunnel and bridge schemes alike to raise capital, even if they did have full political authorization to go ahead with detailed design and construction. This is not to say promotion does not proceed. Yet in order to promote a bridge or any other great project, one does have to have more than a fuzzy idea—one has to have workable conceptual designs, at least. In the case of Lin's peace bridge, the relatively shallow depth of the water—180 feet maximum over the fifty-mile route—places the project well within design and construction experience. Precedent, or at least arguable models, exists in the twenty-four-mile-long bridge across Lake Ponchartrain near New Orleans and the more recently opened seven-mile-long Confederation Bridge across the ice-forming Northumberland Strait, but in the Bering Strait special attention has to be paid to oceangoing shipping, significant ice forces, and protection of bridge traffic from the elements of wind, snow, ice, and cold. Lin's design, for example, has the railroad trains running inside a huge box girder, reminiscent of the scheme used in the Britannia tubular bridge, thus shielding the rolling stock from the elements.

Whether dream bridges will be realized within the foreseeable future depends at least as much on political and economic conditions—and will—as on technical details. In the meantime, bridge watchers will have to content themselves with more down-to-earth projects, such as the replacement of the East Bay crossing of the San Francisco–Oakland Bay Bridge. This is the portion of the bridge that suffered damage in the 1989 earthquake. Subsequent reassessment of the bridge led to a realization that retrofitting the structure to make it earthquake tolerant not only would cost on the order of a billion dollars but also would make the bridge even less aesthetically pleasing, especially compared to its near neighbor across the bay, the Golden Gate. When it was believed that a completely new structure could be had for about the same amount of money, that became the way the Bay Area wished to go. Among the proposals was one employing "an array of horizontally tensioned cables to support a lightweight road deck" and thought to be capable of spanning in excess of two miles. However, the panel of engineers and architects who reviewed the more than twenty design options in the end chose a single-towered, anchorage-free suspension bridge, a design concept superficially not unlike that proposed for other locations in the mid-nineteenth century. Such coming around suggests that even the technical community is not always ready for new ideas or fully sure of how far or how fast it can or wants to go at a given time. This is especially true in the case of great bridges.

And Other Things

Dorton Arena

Not very long ago, I picked up at the Raleigh-Durham airport a prominent and very busy structural engineer who had flown in from Chicago to give a talk at Duke University. As we were driving toward Durham, I asked him if there were any structures in the area that he would like to see during his brief visit. He did not hesitate before naming just one: Dorton Arena. Since we were driving in the opposite direction from Raleigh and the North Carolina State Fairgrounds on which the arena sits, and since we did not have that much time to spare before his talk, I told him we would risk getting caught in traffic if we went there first. Perhaps we could go by after his talk.

Dorton Arena, completed in 1952, was an early stadium-like structure designed to enclose a large, covered, column-free space. It is thus a predecessor of such covered stadiums as the Houston Astrodome, completed in 1965, and the variety of superdomes that so dominate sports-stadium design today. Yet for all of its influence on subsequent design and construction, the significance of the unique Dorton Arena remains relatively unknown outside the structural-engineering and architectural communities. Instead of relying strictly on the primitive principle of compression, in which loads are carried by bearing down on what supports them, as do pyramidal piles, planar walls, circular arches, and domes of stone, the Dorton Arena is a structure with a roof that is supported in tension.

Tension structures carry their loads by resisting being pulled apart. Suspension bridges, with their graceful cables, are very prominent tension structures. Tents are also tension structures, with their fabric stretched over poles the way the bridge cables are slung over towers. Indeed, although not often visualized in this way, a large circus or event tent can appear in profile or silhouette to resemble a suspension bridge or a number of suspension bridges in tandem. Because tensile structures work by resisting being pulled apart, they also have to pull against something; thus, suspension bridges require anchorages, as tents require stakes in the ground, to maintain their configuration. Among the Dorton Arena's unique features is the elimination of any anchorage or stakelike components, making it a more economical and elegant structure. So how does it bear the essential tension?

The roof cables pull against a pair of crossed and inclined concrete parabolic arches that are perhaps the arena's most dominant feature. Like all arches, these work in compression, and the pull of the cables in the plane of each arch is transformed into a compressive force that flows down the legs of the arch into the ground. The structural action of

Structural principles of tension structures

the building has been given the anthropomorphic interpretation of being like two men who lock arms and pull against each other. If they were standing upright, their mutual pulling action would tend to make them fall toward each other. To keep this from happening, each of the men can place his feet behind those of the other as he leans backward and pulls with his arms. In this cross-tied stance, each would only fall backward if he let go of the other. However, the men might also slide on the ground if they did not have their heels dug in or their legs tied together in some way. An analogous action is going on beneath the surface in the Dorton Arena, where the feet of the arches bear against massive abutments and where the crossed legs are tied together with steel cables.

The structural action of the arena as embodied in the human model is considered a brilliant, original, and extremely elegant solution to the ever-present engineering problem of equilibrium. By making the structure's arms out of steel cables, which are very efficient in tension, and making the leaning body out of concrete, which is very efficient in compression, the best properties of both materials are exploited. If the arches were not inclined—if the men were not leaning backward—the structure would not work. For two men to pull against each other while standing upright, an additional structural element would be required. This could take the form of another man lying behind each and holding him back with a rope. Then the structure would be analogous to a suspension bridge, with the backup men acting like anchorages. As beautiful a structure as a suspension bridge can be, its need for anchorages (and the extra cost they entail) puts a blemish on it as a less efficient structure than the Dorton Arena. But how did this elegant structure come to be, and why is it on the state fairgrounds in Raleigh, North Carolina?

The principle of using cables to support a roof, albeit not a permanent one, is said to have been employed in the Roman Colosseum two thousand years ago. Some suspended roofs were evidently used in Russia at the end of the nineteenth century, but steel strips rather than cables were used as the support elements. Suspension-bridge development, especially in the later nineteenth and early twentieth centuries, provided a testing ground for using steel cables in large-scale tension structures. Yet the Dorton Arena's permanent cable-supported roof system was the first one in the world on such a large-scale structure, one measuring the length of a football field in each direction.

The origins of the Dorton Arena lay in the postwar years, when J. Sibley Dorton, a veterinarian turned fair manager, promoted long-range plans "for future expansion of the North Carolina State Fair into a permanent State Exposition on a year-round basis" in the temperate climate of the Piedmont. He envisioned a "modern, well-planned Exhibit Arena and Assembly Building, adequate to accommodate and seat 15,000 people." The building, which was the centerpiece of his vision, would "provide adequate facilities in the amphitheatre for the proper showing and sale of all forms of livestock, as well as shows for automobiles, textile machinery, and every other conceivable type of industrial shows and sales, as well as all forms of sports and athletic events."

The architect chosen to flesh out Dorton's vision was William Henley Deitrick, who was born in 1895 in Virginia. He developed ties to North Carolina by attending Wake Forest College and by marrying Elizabeth Hunter of Raleigh, the city in which he set up his architectural practice in the mid-1920s. He thus had more than two decades of experience designing buildings and planning land use in the state before he received the North Carolina fairgrounds commission.

Since the late 1930s, Deitrick's firm had practiced out of a building dubbed the "ivy tower office," owing to its location in the renovated Raleigh Water Works, with its distinctive octagonal vine-covered granite tower. The Tower—as the building and its offices inevitably came to be known—has been described as "a professional training Mecca for young architects" of the time. Among the young architects that Deitrick engaged periodically as consultants were Matthew Nowicki and his wife, Stanislawa. Matthew Nowicki had been born in 1910 in Russia but received his architectural training in Poland, where he began to practice. Among his commissions was a sports center in Warsaw. He also had ties to the United States, having studied as a boy at the Art Institute of Chicago, having later served as cultural attaché to the Polish consulate in that city, and having been the Polish representative to the Committee for the United Nations Building in New York, a building for which he also served as a consulting architect.

In 1948, Nowicki joined the School of Design at what was then North Carolina State College in Raleigh, serving as acting head of the department of architecture. He spent a summer at Cranbrook Academy, in Michigan, as a consultant to Eero Saarinen on a planning study for Brandeis University. Back in Raleigh, he served as a consultant on several Deitrick projects, working on the interior of the Carolina Country Club, a state art and history museum, and preliminary designs for the state fairgrounds—including an arena, which, during its development, came to be referred to as the livestock-judging pavilion. Nowicki also collaborated with and consulted for architects in California and New York on both domestic and international projects. Shortly after the arena was commissioned, Nowicki, who was returning from India, where he had consulted on the design of the layout for a new capital city in Punjab Province, was killed in a plane crash in Egypt. One critic called the untimely death of the forty-year-old "a catastrophe to architecture."

The sketches for the arena that Nowicki left behind show clearly the structural concept for the building that was eventually realized as Dorton Arena. One sketch for an early scheme shows an uncovered bowl-like structure, while another shows the bowl covered with a sagging roof supported by inclined arches resting on battered columns. Other sketches show the peripheral columns upright, as the structure was eventually built. However, sketches of the kind Nowicki left are not sufficient for constructing a building, especially one so innovative. There remained considerable work to fill in the details of the scheme and choose the exact materials and sizes for the structural bones and protective skin.

The office of William Deitrick took over the supervision of the detailed work of design and construction, with Stanislawa Nowicki apparently providing considerable insight into the creative intentions that her late husband had shared with her. It was essential to involve a consulting structural engineer in the development of such a novel design, and Deitrick chose to work with the New York firm Severud-Elstad-Krueger. Fred N. Severud, who was described as a "creative engineer" and was a close friend of Deitrick, served as structural consultant. Severud developed the structural system that made the pavilion stand, and he would later extend the principles of the building's long-span cable-net roof to structures such as the Yale Ice Hockey Rink in New Haven, the Reception Building at Dulles Airport, and the circular-roofed Madison Square

Garden in New York. Severud Associates, as the firm is now known, was also the engineering firm for such projects as the Gateway Arch in St. Louis, the Denver International Airport, and the Solomon R. Guggenheim Museum in New York.

Whatever changes mechanical principles would force in the details of Nowicki's conceptual design for the structure, Deitrick was insistent that they be held to a minimum, so that the completed building look as much as possible "as Matthew would have wanted it." This is not to say that there was no room for creative engineering and construction, for Matthew Nowicki did not leave much beyond his sketches. Among the practical details that remained to be worked out were how the building, once designed in detail, would be constructed. This task fell to William Muirhead, a contractor from Durham.

Most structures are built from the ground up, of course, and so the livestock-judging pavilion began with foundation work: footings for the columns, walls for the basement, and the arena floor. Like the men in the anthropomorphic model, the two leaning parabolas, which the weight of the roof would tend to make slide past each other, had to be able to push against something to keep that from happening. Thus, underground concrete abutments were constructed to take the thrust, and these abutments were connected by tunnels through which steel cables could pass to connect the bases of the parabolas to each other. The cables further check any tendency for the bases to slide apart, which would bring the parabolas closer together and let the roof sag below its desired profile. The steel columns that support the concrete parabolas were erected next, thus providing perches onto which a wooden form could be constructed to hold the concrete until it set. (A special mix of concrete was used so that it would stay put until set and not run down the twenty-two-degree incline of the parabolic legs.) Among the many elegant details of the design is the use of the outside columns also as mullions to hold the window frames and glazing in place. The columns, spaced six feet apart, thus serve an efficient, dual purpose.

The roof was put in place by first installing steel cables to span the space between the backward-leaning parabolic arches. To insure that connecting hardware was located in the right places to receive the ends of the cables, careful surveying work had to be done to transfer precise locations into the concrete formwork. The surveyor responsible for overseeing that everything was where it should be was John R. Gove of Chapel Hill, the third of the North Carolina cities that establish the vertices of the state's Research Triangle. In an early use of a digital computer, the locations for the cable sockets were precisely calculated from equations describing the arch. Each cable had to be accurately sized beforehand, so that it hung with the proper sag. (The finished roof would have a maximum sag of about thirty-one feet over the three-hundred-foot span.) There are actually two sets of cables in the roof, at right angles to each other, forming the so-called cable-net.

It was originally thought that the roof would be covered in some kind of fabric, but, as often happens in building, what was available when construction bids were submitted was considerably more expensive than an alternative. To save money, the cable-net was thus covered with corrugated steel panels, on top of which was placed insulation, and on top of that conventional waterproof roofing materials. Among the concerns about the roof, no matter its composition, was that it would flutter in the wind. (The upward suction of the wind was calculated to be as high as sixteen pounds per square foot, whereas the weight of the roof that resisted uplift was only six pounds per square foot.) To prevent the roof from moving upward in the wind, guy cables were installed between cable-intersection points on the interior of the roof and the structural columns around the periphery. The soundness of the roof design was tested when Hurricane Hazel passed almost directly overhead in 1954. The roof weathered the storm, with winds estimated to have gusted to around one hundred miles per hour.

When completed in 1952, the livestock-judging pavilion was hailed as a unique structure enclosing a unique space. Indeed, it was the latter that Nowicki had set out to achieve, and in the course of developing his scheme for it he had come up with the former. Nowicki wished to give every spectator not only an unobstructed view of the arena floor, which was large enough to host a standard horse show, but also a sense of openness. Whereas those sitting in the topmost seats in most covered stadiums of the time had to watch their heads, lest they bump them on the roof, Nowicki's saddle-shaped roof gave even the uppermost spectator a sense of openness not significantly different from that experienced by someone sitting near the arena floor. The cable-net roof was

Dorton Arena

described as "the exact reverse of a dome," with the roof's upward curving ends allowing a maximum amount of light to come in through the windows. According to one critic, the interior was lit in a way that "marked a new epoch in architecture." The pavilion, which was built at a cost of $1.5 million, was designed to hold almost five thousand spectators in permanent seats. Another four thousand could be accommodated if chairs were set up on the arena floor.

Among the honors that the structure soon received were the Engineering Gold Medal of the Architectural League of New York and the First Honor Award of the American Institute of Architects for 1953. In 1957, it made the AIA's list of ten buildings expected to exert the most influence on design in the next century. The Museum of Modern Art in New York was among the museums in which a model of the building was exhibited. The immediate, enthusiastic, and unqualified architectural success of the livestock-judging pavilion led one contemporary critic to ask, "Why, in this land of the engineer, are there so few frank expressions of integrated engineering that create dramatic architecture?" It was certainly a valid question then, and it remains a valid one now.

Dramatic architecture, like dramatic engineering, begins with a creative idea for solving an old problem in a new way. There were plenty of unimaginative livestock-judging arenas located in state fairgrounds around the country, and there were numerous covered riding arenas and sports stadiums that could have served as models for what was needed in Raleigh. Duke University's Cameron Indoor Stadium, fewer than twenty miles away in Durham, was just one example right in the area. Indeed, many architectural and engineering problems are solved by adapting existing solutions with minor modifications, often without regard to the unique needs or opportunities of a new site. Matthew Nowicki, by thinking about the problem anew, perhaps using his European experience and training to reach beyond examples to principles, was able to rise to the occasion and propose a truly imaginative scheme. And his inspiration was well served by Stanislawa Nowicki, his widow and collaborator; his friend William Deitrick, the architect; and his friend Fred Severud, the engineer. With the blessing of J. S. Dorton, representing the client, and the sympathetic construction skills of William Muirhead, the contractor, Nowicki's dream was realized. The formula for dramatic engineering and architectural achievement is on one level simple, then: Just recognize a brilliant idea when it occurs, and preserve it through the long and arduous process of making it a reality.

When it opened, the fairgrounds pavilion that was built "to serve agriculture, industry, and commerce" was officially and unpretentiously named the State Fair Arena. Unofficially and unappreciatively, it was referred to as the Cow Palace. While under construction, it had been called "a flying saucer anchored to a glass platform" and shortly after its opening a "parabolic pavilion." But from the start it was also recognized among professional architects and engineers as "the most important building in America today."

In 1961, at the opening ceremonies of that year's fair, the building was renamed the J. S. Dorton Arena, in recognition of the longtime manager of the North Carolina State Fair, who had recently died. The building was placed on the National Register of Historic Places in 1976, a considerable honor for a structure barely a quarter century old. In

2002, just as it reached the half-century mark, it was named by the American Society of Civil Engineers as a National Historic Civil Engineering Landmark, a distinction that cannot be applied to anything less than fifty years old. A plaque declaring the structure's landmark status was unveiled at the opening day of the 2002 fair.

No matter how much recognition the Dorton Arena has received, it continues to function as a state fairgrounds building. The fair can occupy the arena only a couple of weeks a year, however, and so the building is available to rent the rest of the time—at a cost of $1,400 per day or 10 percent of the gross ticket sales, whichever is greater. Among the high-profile events that the arena has hosted have been comedy shows and rock concerts. It has also been the site of farm shows, home shows, ice-hockey games, basketball games, high-school proms, wrestling matches, and circuses.

In spite of its humble beginnings and unpretentious uses, the Dorton Arena remains "a familiar architectural landmark, studied and hailed as a masterfully unique design, nationally and internationally." Unfortunately, getting my visitor to the airport to catch his return flight did not allow us to go to Raleigh to view the structure. He went back to Chicago not having seen one of the most significant engineering and architectural structures in the Research Triangle area and in the world.

Bilbao

The Guggenheim Museum in Bilbao, Spain, has been called one of the most complex and significant buildings of the late twentieth century. Engineers marvel at its structure, and architects flock to see its texture and space. Indeed, even before Museo Guggenheim Bilbao—or just Bilbao for short—was completely finished, the building was the setting for the awarding of the 1997 Pritzker Prize, the highest honor in the architectural profession. This is all quite an achievement for a structure with a number of stories, although perhaps difficult to define but certainly no more than about three, that are nowhere near in number those of the soaring skyscrapers that have generally been the focus of architectural and structural attention throughout the twentieth century.

How such a world-class piece of architecture and engineering came to be located in an industrial port city that had seen better times is a tale that began in the 1980s. Local boosters wanted to bring to the provincial Basque capital of Bilbao in northern Spain some of the attention, respect, and tourism that were expected to come to Seville, which had commissioned several significant bridges for its 1992 universal exposition or world's fair, and Barcelona, which was to host the Olympic Games that same year. A major museum, in which the great art of the world could be displayed, was one route that Bilbao's movers and shakers saw to distinguish their city and, not incidentally, attract tourists to it.

After it was decided that the Basque administration would finance the project and that New York's Guggenheim Museum would manage the new museum and display some of its collection there, it was time to provide a building. The Guggenheim's landmark helical building, designed by Frank Lloyd Wright, certainly has a cachet invaluable to the museum and to New York City, for many people come to see the structure itself, a piece of art in its own right, as much as to view the art it contains. If Frank Lloyd Wright could do that for a museum in New York, perhaps a late-twentieth-century architect could do the same for one in Bilbao and thus put it on the cultural map. Such is the motivation for not a little monumental architecture, from museums to skyscrapers. In the earliest stages of planning, however, such clarity of purpose is not always evident, and the initial plan had been to house the Spanish museum in a "venerable old wine warehouse" downtown.

In the 1980s, Bilbao's image was that of a working-class city that had once been the center of Spain's steel industry, located there largely because of nearby iron-ore deposits. Steelmaking had led to shipbuilding, which in turn encouraged banking and insurance companies to flourish in the city. But that was in the past, and Bilbao in the late twentieth century needed to be revitalized. World-class architects were to be commissioned to help with that revitalization. Sir Norman Foster was chosen to design a subway system that relieved traffic. Cesar Pelli was given the opportunity to redevelop a riverbank that had been the location of warehouses, factories, and a railroad depot. And Santiago Calatrava, whose Alamillo Bridge in Seville had already become well-known, was to build a distinctive pedestrian bridge and also design a new airport for the city.

In the case of the museum, the Basque and Guggenheim leaders invited three architects to participate in a 1991 design competition: Arata Isozaki & Associates of Japan, Coop Himmelblau of Austria, and Frank O. Gehry & Associates of Santa Monica, California. Both Himmelblau and Gehry argued that the museum should be located not downtown but on a riverfront site, and it was Gehry's design for an unconventional structure in an abandoned industrial area at a bend in the Nervion River that won the competition. The site, formerly occupied by a factory and a parking lot, was also intersected by the Puente de la Salve, a cable-stayed bridge that carries a main traffic artery into Bilbao. Rather than seeing the bridge as an obstacle, Gehry saw it as something to work with, and he designed one of the museum's galleries—a 450-foot-long-by-80-foot-wide columnless space suitable for accommodating the truly large-scale multimedia works that are so favored by contemporary artists—to pass under the bridge and effectively incorporate it into the museum's design. (Another gallery, designed especially to house *Guernica,* Picasso's large painting commemorating the bombing by the Nazis of the nearby town of that name, has yet to see the function of its space realized. Bilbao's request to Madrid that the painting be loaned was turned down because *Guernica* was said to be too fragile to travel. Basque observers saw the decision as motivated by politics.)

Gehry's international reputation is that of an architect who uses distinctive forms and new materials. He has designed private residences as well as concert halls, and for his body of work he had been awarded the Pritzker Prize in 1989. His own home started out as "a small pink two-story house in a middle-class neighborhood," which he redesigned and expanded by building a new structure around three sides of it, thus preserving much of the original and its cultural evocations. The materials Gehry used on his house included corrugated metal, plywood, and chain-link fencing, and it was not everyone's idea of improving the neighborhood. But Gehry's unconventional genius has also been manifested in his design for the Walt Disney Concert Hall for Los Angeles and, of course, the Bilbao museum, which the Norwegian architect Sverre Fehn, who received the 1997 Pritzker Prize, described as "an instant sketch that has been realized."

How an architect's sketch, instant or otherwise, is realized is often no easy task. In the case of Bilbao, as with many of Gehry's designs, the walls were typically neither vertical nor flat, and so devising a structural frame to hold up the sculptural building was no ordinary feat of engineering. However, the situation is not unique to the late twentieth century or to Gehry's buildings. When the sculptor Frederic-Auguste Bartholdi designed the Statue of Liberty to be a gift from France to the United States on the occasion of its centennial, he did not himself deal with how to hold steady in the wind of New York Harbor the beaten copper sheets that would be shipped across the ocean to make up the 151-foot-tall statue. That job eventually fell to the bridge engineer Gustave Eiffel, who devised a structural frame of wrought iron to which the copper shell of the statue could be attached and by which it would be supported. A mistake made during the erection of the ironwork led in time to a weakened arm on the statue, a condition that prevented tourists from continuing to climb into its torch. In addition, when the statue was closed for maintenance and repair in preparation for America's bicentennial, much work had to be done on the connection points between the copper skin and the iron frame, where an electrochemical reaction between the dissimilar metals had led to considerable corrosion.

Another instant sketch of a sculptural structure that was realized only after considerable time and money were spent is the Sydney Opera House. When the city of Sydney, Australia, decided that it wanted a new and distinctive opera house, it launched in 1955 an open international competition. The winning design surfaced in the form of striking drawings of soaring shell-like structures submitted by Jorn Utzon, then a little-known Danish architect. Realizing the design in concrete was no easy task, however, and the original idea of casting the shells in place was abandoned because it would have been prohibitively expensive to design and erect the large and complex forms that would have been needed. Utzon's next idea was to cast individual rib pieces and join them together to make the shells, using ceramic tiles to fill in spaces between the ribs. Debates over construction methods and other disagreements with a new government led to Utzon's resignation in 1966. The project was completed by a team of Australian architects in 1973, fourteen years after construction began, and at a final cost that was almost fifteen times the original estimate. Although the Sydney Opera House has become a symbol of the city, it remains a disappointment to opera lovers, because large-scale operas cannot be adequately staged in it.

In contrast, the Bilbao museum was completed pretty much on schedule and about 10 percent under its hundred-million-dollar budget. A principal reason for the remarkable achievement is the use of computers, in particular the use of computer-aided design and computer-aided manufacturing, known as CAD-CAM. Bilbao started, as do many of Gehry's designs, with numerous small-scale models in wood, paper, and cardboard. To transfer these three-dimensional forms into digital data, an electronic probe was used to measure the exact coordinates of points on the surface of the physical model. These were converted into a computer model that could then be manipulated with software developed first for the French aircraft industry. This same software, known as CATIA (standing for Computer Aided Three-dimensional Interactive Application), was used to design and manufacture the Boeing 777 with virtually no need to rely on conventional paper drawings. CATIA was ideally suited for use on Bilbao, because the proliferation of curved surfaces that were to be clad in metal made the structure more like that of an aircraft fuselage than that of an earthbound building.

With a computer model of the building created, Gehry could manipulate it electronically on the computer screen to make changes that he felt were appropriate in the building-as-sculpture. The computer could then be used to control machines that were capable of generating new physical models, which in turn could be compared with Gehry's own models and concepts. In this way, the final form of the building design was arrived at and, most important, the same design data and computer software were then available for assisting in designing the structure and operating the steel-cutting and steel-forming machinery. Although the building's metallic cladding would not itself be formed via computer, the extensive database would be employed to track the location and arrangement of the numerous panels.

Before any cladding—conventional or otherwise—could be attached to the building, however, it needed an appropriate structural skeleton to hold it up. The conventional wisdom was that a building of such complex curves should be cast in concrete, a totally plastic material but one that would have required the use of elaborate formwork to make the mold into which it would be poured and held in place until cured. (This is the way the Guggenheim Museum in New York was constructed and the way Utzon imagined that the Sydney Opera House might have been fashioned.) The structural design of the Bilbao Guggenheim was to be the work of engineers in the Chicago office of Skidmore, Owings & Merrill, out of which had come, in addition to concrete structures, such innovative steel buildings as the John Hancock Center and the Sears Tower. SOM engineers Hal Iyengar, Lawrence Novak, Robert Sinn, and

Underlying steel structure of Bilbao Guggenheim Museum

John Zils came up with a concept that convinced Gehry's design team that the museum could be framed in steel. Again, it was the computer modeling and software that made such a decision possible.

No two pieces of steel are said to be exactly the same, but there are essentially only three kinds of members used in Bilbao's structural frame. The typical vertical member is a wide-flanged section, the kind of steel piece that often forms the vertical or column elements in a warehouse-style superstore or the exposed structure of an airport such as that of the new terminal at the Ronald Reagan Washington National Airport. The typical horizontal member in the Bilbao museum is a square tube, and the typical diagonal is a round pipe. The structure is based on a three-by-three-meter grid, and at nodes where the three types of members join steel plates allowed them to be fitted together at the different angles needed to produce the overall shape that Gehry approved. As it did with the cladding that would be attached to the finished frame, the use of the CATIA software enabled the individual steel members to be designed and manufactured with relative ease. Although welding was at first expected to be used to join the six members at a typical node, in the end bolting was decided upon as quicker and more easily controlled. Since all joint designs were computer generated and all joint members were computer manufactured, there was little problem with bolt holes not lining up.

Given the elaborately curving shape of the building's facade, the steel frame is remarkably regular in its design. However, unlike a more conventional building, in which most of the rectangular spaces framed by vertical and horizontal members are open, with the entire structure being stiffened against sway by diagonals located only here and there, in Bilbao every one of the rectangular spaces is bisected by a diagonal pipe. During construction, this gave the smallest part of the incomplete structure a strength and stiffness that enabled it to be erected without temporary scaffolding, even where the walls leaned outward, thus turning the constantly curving geometry into a structural asset. Indeed, the frame of the Bilbao museum is more like a three-dimensional steel shell than the rectilinear jungle gym that holds up most conventional steel office buildings. When the steel structure was completed early in 1996, it was ready to be clad in metal and stone.

Gehry had long admired the "soft, buttery look" of titanium, and he had wanted to use it to clad a large structure. But the price of the metal—put to such lofty uses as the heads of premium golf clubs and the landing gear of aircraft—had been prohibitively high for a building that did not need titanium's strength. However, a combination of events lowered the price of titanium below that of stainless steel just as the Bilbao building was to go out for bids. That, along with the fact that the envelope of rolling technology could be pushed to manufacture sheets of titanium only 0.38 millimeters thick, made it an appropriate cladding material after all.

Not all of the building is clad in titanium, however, for Spanish limestone was used on a considerable part of the structure. To cut the limestone with the curved surfaces needed to match Gehry's design, a steel milling machine was modified to work on stone. It is believed to be the world's first numerically controlled stonecutter capable of producing surfaces with different curvatures in two different directions.

Bilbao facade

Furthermore, one end of the building consists of a tower of structural steel that Gehry deliberately left unclad, thus exposing the underlying structure and giving the museum the look of a work in progress.

The finished building, believed to be "the only building in the western world with a significant amount of titanium," has been described as "an eruption of powerful forms held in check by a cloak of titanium—a metal that does not so much reflect the light as capture it and send it back altered, in a range of tones from bright silver to deep, purplish gray." According to the critic Calvin Tomkins, "From across the Nervion River it looks like a fantastic dream ship, all sails full, sweeping upstream into the center of somber nineteenth-century Bilbao; walk two blocks further and it starts to resemble some prehistoric beast, advancing with one huge leg and foot toward the water." Although Tomkins may have seen the museum as a beast moving toward the Nervion, Gehry himself said that the idea of the building looking like a boat was his "response to the river." Whatever it may remind viewers of, however, the museum building is considered by virtually everyone as a piece of art in its own right, and Philip Johnson, who is known as the dean of American architects, is reported to have declared the Bilbao Guggenheim "the greatest building of this generation."

Santiago Calatrava

The conception and design of most monumental structures involve a collaboration of architect and engineer. However, except in the case of bridges, which are often considered to be pure engineering, it is usually the architect who gets credit for the creative idea. According to the conventional wisdom—and most often in actuality—it is the architect who first sketches grand plans and facades and the engineer who calculates what beams and columns are needed to make the artistic concept work. Even for plans never realized, architectural drawings can take on lives of their own and be the subjects of art shows, but the blueprints of engineers are seldom seen outside the design office or construction site. Engineers curse architects who conceive buildings that require extraordinary contortions of heavy concrete and steelwork to execute; architects look down on engineers whom they accuse of lacking the structural creativity to solve unusual problems or the willingness to let their imaginations soar.

As with all caricatures of groups of people and professions, this oversimplified view has both elements of truth and many exceptions. Sydney Harbour provides an excellent juxtaposition of contrary examples. Jorn Utzon's original sketches for the Sydney Opera House so captivated the competition judges that little consideration appears to have been given to constructability. It took the innovative engineer Peter Rice, who would later have so much to do with the exposed structure of the Centre Pompidou in Paris, six years to work out the practical matters of the shells in Sydney. And still, within about fifteen years of its completion, the condition of the structure had deteriorated to such an extent that it was evaluated to need repairs costing about as much as the structure itself.

In contrast to the opera house, the neighboring landmark Sydney Harbour Bridge, like most major bridges, began not with an architect's sketch but with an engineer's concept of the structural form that would be most appropriate for the site. In fact, a design was put forth as early as 1857, and many others followed, but no scheme was agreed upon for more than fifty years. In 1911, John Bradfield, principal design engineer of the New South Wales Public Works Department, began to work on designs that would be buildable and functional as well as distinctive. His first plan was to erect a nondescript and faddish cantilever, but this was finally changed when he visited America and saw Gustav Lindenthal's design for the Hell Gate Bridge in New York City. The cantilever plan was tabled, and Bradfield had the Sydney Harbour Bridge redesigned as an arch, which yielded a scaled-up version of the Hell Gate with some improved structural details. The contribution of architects was by and large limited to the stone pylons that frame the bridge proper, which opened in 1932.

Some of the most innovative and effective works of architecture and engineering have been true collaborations, and two of the most striking examples stand in Chicago. Both the John Hancock Center, completed in 1969, and the Sears Tower, finished in 1974, were made possible by the structural engineer Fazlur Khan and the architect Bruce Graham working together in the firm of Skidmore, Owings & Merrill. For such tall buildings to use steel economically and yet stand steady against the winds of Chicago, their structural bones and architectural skin had to evolve and work together, with engineer and architect influencing each other's concepts of form and function. Rather than presenting to the observer mere facades hung from structural skeletons, the Hancock and Sears towers are in fact tall, slender structural tubes that themselves shape the facades. There have been numerous other successful collaborations among engineers and architects, especially in the design of tall buildings, but some of the most talked-about ground-level structures today are coming not from teams but from the mind of a single individual who himself is both architect and engineer.

Santiago Calatrava Valls was born in Valencia, Spain, in 1951. He studied art before entering Escuela Técnica Superior de Arquitectura de Valencia to become an architect, and he subsequently did graduate work in urban studies. According to Calatrava's own account, during his studies in Valencia he developed a "great admiration for the work of engineers" and recognized the need for "more technical knowledge to do the things I wanted to do." In 1975, Calatrava left his hometown for Zurich to study civil engineering at the Eidgenössische Technische Hochschule (the famous ETH, or Swiss Federal Institute of Technology). He saw engineering as "the art of the possible" and ultimately received a doctorate from ETH with a thesis on the foldability of space frames. He also served as an assistant in the Institute for Building Statics and Construction and for Aerodynamics and Lightweight Constructions before leaving ETH in 1981, at the age of thirty, to found his own architectural and civil-engineering practice in Zurich. The firm, which would open an office in Paris in 1989, is formally known as Calatrava Valls S.A. Unlike Ludwig Mies, whose given name tended to be dropped after he added his mother's surname, van der Rohe, in the Spanish tradition Santiago Calatrava has tended to drop the matronymic in most contexts.

For a less deliberate person, the apparent flitting among art and architecture and urban studies and civil engineering and technical science might presage an undirected life of frustration and dilettantism. For Calatrava, however, it appears at least in retrospect to have been the carefully planned, if unorthodox, curriculum of an individual determined to be a unique combination of architect and engineer. It must have been daring even for Calatrava to strike out into private practice when he did, for before 1981 he had designed only a single highway bridge and a single exhibition hall, neither of which was built. However, for a confident, if budding, architect/engineer like the young Calatrava, with dreams of designing great structures that were also works of art, Zurich might have seemed to be a city of opportunity with a tradition of structural genius. It was in Zurich, after all, where the great Swiss bridge designer Robert Maillart began his career, and it was to Zurich that the successful bridge designer Christian Menn, who had been so influenced by Maillart, returned to become professor of structural engineering at ETH—his, Maillart's, and Calatrava's alma mater.

Even without any of his early design projects being realized, Calatrava was quickly able to make a name for himself because of the European practice of holding design competitions for proposed civic works. Whereas in the United States choices among competing designs are often made on the basis of cost, in Europe selections tend to be more influenced by less quantifiable questions of aesthetics and symbolism. Even if one's design does not win in a European-style design competition, the occasion provides a venue for displaying one's talent, vision, and promise. Calatrava entered a half-dozen competitions between 1979 and 1982, for projects ranging from highway bridges and exhibition halls to factories, warehouses, and libraries. It was not until 1983 that he received a commission to bring any of his designs to fruition, but then his practice took off, with commissions coming to him even without competitions.

Calatrava's designs—even when they exist only as sketches, drawings, or models—are striking works of art. His buildings and train stations are rich in sculptural form and detail, and his bridges are dynamic assemblages of arches and cables with dramatic approaches and spaces for people to use and enjoy. By the mid-1980s, exhibitions dealing with Calatrava's work began to appear in Swiss art galleries and museums, and by the end of the decade an exhibition was organized to travel to America. Awards also began to come Calatrava's way—from art, architecture, and engineering associations alike—including gold medals from both the French Academy of Architecture and the British Institution of Structural Engineers.

Among his earliest completed designs is the Bach de Roda–Felipe II Bridge, completed in 1987 in Barcelona, a structure in which two outside steel arches supporting walkways lean over pedestrians and against two more conventionally placed vertical arches on either side of a central roadway. The cables from the arches to the bridge deck enclose pedestrians in a broad space that gives an almost shell-like feel to what might in lesser designs have been a narrow path across a busy bridge. Perhaps Calatrava's most famous bridge to date is his Alamillo Bridge, built for the 1992 expo in Seville. This dramatic cable-stayed structure has a single 142-meter-high pylon angled back from the main span. Between the pylon and the two-hundred-meter-span bridge deck stretch the steel cables in a harplike arrangement.

None of Calatrava's bridges is particularly long or daring in span, and his first British project that went beyond the drawing board was a sixty-two-meter-long footbridge, the design of which was unveiled in 1993. Trinity Bridge, which spans the River Irwell between Salford and Manchester, is also supported by a single, inclined pylon (sixty meters high), but the cables are in a much more expansive three-dimensional arrangement, and the roadway splits on the Salford side to make the bridge accessible to the handicapped via two gracefully curving ramps. The bridge was designed to be the centerpiece of a riverside redevelopment project.

Although by the early 1990s bridges may have been considered the mainstays of his practice, Calatrava also established his reputation as a designer of spectacular railway stations and other public spaces. His Stadelhofen Railway Station in Zurich, completed in 1990, expanded the capacity of the rail line on a complicated curved urban site from two tracks to three, added an underground transfer gallery lined with shops hidden between massive but graceful "zoomorphic" concrete pillars, and let daylight into it all to brighten what might otherwise have been a dark and heavy space indeed. The Satolas Airport Railway Station, fifteen kilometers east of Lyon, France, was completed in 1994 and uses concrete—"the most noble" of construction materials, according to Calatrava—in a much lighter and dynamic way, in keeping with the high-speed French TGV trains that pull into and out of the station. Above ground, Calatrava's five-hundred-meter-long steel, aluminum, and glass building that straddles the railroad tracks gives the traveler the impression of being inside the skeleton of some prehistoric giant.

Calatrava, who has said that "a building is a sculpture you walk into," had his first significant opportunity in the United States in 1991, when he was invited to participate in a design competition for the unfinished Cathedral of St. John the Divine in New York City. His controversial concept of a glass-topped, towering transept was for a long while not fully accepted or rejected, and its future remains uncertain.

The first Calatrava-designed building/sculpture realized in the United States is the seventy-five-million-dollar addition to the Milwaukee Art Museum, the original part of which was designed by Eero Saarinen as a war memorial. The Calatrava addition, which was expected to bring renewed attention and visibility to the museum, was selected from designs submitted by almost seventy architects from around the world. Calatrava's original concept called for a main pavilion with a louvered roof that in the closed position looked not unlike a giant wigwam. Opened, it was said to look like an enormous seagull with outspread wings. Some saw the fully unfolded motorized roof as the flukes of a great whale—something unexpected in nearby Lake Michigan.

Addition to Milwaukee Art Museum

The Calatrava addition, which opened in 2001, was extremely well received in Milwaukee and around the world. Everything about the building—from the inclined steel framing the underground parking garage to the long arched galleries to the towering glass-and-steel truncated cone of the entrance lobby and ceremonial space—won high praise from museumgoers and architectural critics alike. But it was the two highly visible external components of the design that came in for special praise. The 250-foot-long cable-stayed bridge with Calatrava's signature inclined mast ties the art museum into Milwaukee's downtown and provides an elegant entrance to the new space and structure.

But most praise was reserved for the kinetic sculpture atop the building, which was designed also to serve the practical purpose of keeping the reception hall from overheating in bright sunlight.

The movable sunscreen—now more commonly referred to by the French term, *brise soleil*—with its 217-foot open wingspread has been described as being the size of a 747 jumbo jet. In fact, the hundred-ton folding screen proved to be a fabrication and construction challenge that threatened to delay the reopening of the museum. Calatrava's initial preference was to use aluminum for the seventy-two tubular fins that could hardly be called feathers, but engineers involved in the construction project argued for using a carbon-fiber composite material that would be lighter, stronger, and more durable. However, well into the project it was evident that using the aerospace-grade material specified by the engineers would be too expensive and take too long to fabricate. The design proved to be "more esoteric than anticipated," and so Calatrava fell back on the workhorse structural material of steel and had the fins made in Spain and flown to Milwaukee in time for the last-minute installation. It takes about four minutes for the *brise soleil* to open and close, which happens on schedule every day except when the wind exceeds about twenty-four miles per hour. When that wind speed is sensed and the sunscreen is open, hydraulic cylinders will close it into position around the glass roof, to "sheathe it like a second skin."

The addition to the art museum, for which the Milwaukee firm of Graef, Anhalt, Schloemer & Associates was the structural engineer, was Calatrava's first major project in the United States, but others were quick to follow. Among other American commissions were a sundial footbridge in Redding, California; a cathedral in Oakland, California; a terminal for the Dallas–Fort Worth airport's people mover; a group of bridges over and a parkway beside the Trinity River in Dallas; an esplanade and bridge to expand Grant Park on Chicago's lakefront; and, what will be perhaps most closely watched and reported upon, the transportation terminal at the World Trade Center site in New York City. Before long, Calatrava structures may be as widely distributed and recognizable throughout America as Frank Gehry buildings have become.

Not surprisingly, such a daring architect/engineer as Santiago Calatrava has his detractors as well as his champions. In the latter camp tend to be art and architecture critics, who see him as "the most exciting shaper of public space . . . today." His bridges and stations, places where commuters and travelers can be so frustrated among rush-hour traffic jams and crowds, come in for especial praise. According to Boyd Tonkin, writing in the *New Statesman and Society* on the occasion of Calatrava's visit to England in 1994, he "redeems the dead space, and dead time, of transit. To get somewhere, we tolerate passing through nowhere. His constructions help to turn those nowheres into some-wheres." The dramatic canted columns and the welcoming vibrant open-work spaces in his structures of transit have become trademarks of Calatrava, making his creations immediately recognizable as his. That is not to say that there is a sameness to his work, for each new design adds new surprises to his growing oeuvre.

Some engineers have criticized Calatrava's designs because they do not make efficient use of materials. For example, to keep the longest cable in the Alamillo Bridge from having an unsightly sag due to its own great weight, it was stressed by a concrete counterweight under the roadway to a greater degree than structurally necessary. Another criti-cism of Calatrava's designs is their constructability—or, rather, their lack of constructability. Most cable-stayed bridges are designed to be self-supporting even when only partially completed, but the unusual design of the Alamillo Bridge did not allow that. Such complications necessarily add to the cost of projects; the final price of the Alamillo Bridge was reportedly twice the initial estimate. It is this aspect of Cala-trava's work that may ultimately frustrate him from realizing one of his stated goals: "I want to win back engineering objects like bridges for architecture."

Some of the criticism of Calatrava by engineers may stem from the fact that his broad educational background has frequently been con-trasted to the narrow specialized training supposedly received by so many engineers. According to Anthony Webster, writing in *Architec-tural Review,* the predominantly technical education of engineers "lim-its their ability to tackle many larger design issues." Furthermore,

[t]his system encourages engineers to develop only very limited design skills, and has resulted in today's ironic situation in which the professionals entrusted with the creation of a significant portion of our built environment are unprepared to consider its compositional

and programmatic aspects. By contrast, Calatrava's unusually broad formal training (as both architect and engineer) fosters a design approach incorporating technology, aesthetics and space-making.

Whatever may be said of or about it, Calatrava's approach to design has captured the attention of architects and engineers and critics alike. In a field where there are so many collaborative relationships between the professions, Calatrava is almost unique as a lone designer who seems to feel comfortable in both camps and who has the talents and self-confidence to combine their skills and sensitivities in his striking and effective designs for civic structures. Even though a large portion of the designs he is noted for still exist only on paper, in computer-generated images, or in striking tabletop models, Calatrava has become the turn-of-the-century architect-engineer to watch well into the new millennium.

Fazlur Khan

The street sign on the northwest corner of the intersection of West Jackson Boulevard and South Franklin Street in Chicago designates the location "Fazlur R. Khan Way." Who was Fazlur Khan, why is a street corner in Chicago named after him, and what is the significance of these questions for engineering?

Fazlur Rahman Khan was born in 1929 in Dacca, in what was then East Bengal, India. (In 1947 it became part of East Pakistan; today, its name spelled Dhaka, the city is the capital of Bangladesh.) According to one account, when young Fazlur discussed with his father, who was a mathematician and educator, whether he should study physics or engineering, the elder Khan offered his opinion that "engineering would be better because it demanded discipline." Fazlur Khan later admitted that his decision was also influenced by the fact that "at the time the engineering challenges in India and East Pakistan looked more promising that the challenges in physics." In any case, his comment that he "liked physics because of its mystical and abstract aspects, but I was good in mathematics and always tinkering with something mechanical," expressed an ambivalence about what to study that has been felt by countless engineering students around the world and will no doubt continue to be felt in generations to come.

Khan attended the University of Dhaka and received a B.S. in civil engineering in 1950. After graduation, he worked for two years as an assistant engineer for the highway department and taught at the University of Calcutta. Qualifying for a Fulbright scholarship in 1952 and also winning a Pakistani government scholarship, he went to the University of Illinois at Urbana, where he earned M.S. degrees both in civil engineering and in theoretical and applied mechanics and, in 1955, a Ph.D. in structural engineering.

Fazlur Khan

Like many a foreign graduate student in engineering today, Khan had as a condition of the financial support of his education an obligation to return to his home country within a year and a half of completing his last degree. But, also like many a foreign student, he wished first to gain as much engineering experience as he could in America. He "was considering offers from top engineering firms across the country," when by chance he met a friend who worked at Skidmore, Owings & Merrill, a firm noted for its integration of architecture and structural engineering. Hearing of an exciting project being worked on at SOM, Khan impulsively went to the firm unannounced, "had an interview, got an offer and immediately accepted at a salary considerably less than any other offer" that he had. Khan took the job at SOM because, whereas the other firms had told him he would "start on some beam analysis," at SOM he was given immediate responsibility, with his first project involving "complete charge of seven prestressed [highway and railway] bridges at the U.S. Air Force Academy in Colorado." Recalling the incident twenty-five years later, Khan said, "Without such opportunities life doesn't move. You have to grab them."

Khan returned to East Pakistan in 1957, working as a consulting engineer, as director of the country's Building Research Center, and later as technical adviser to the Chief Engineer's Office of the Karachi Development Authority. Although the latter positions provided considerable status, the administrative workload kept him from doing the design work that he wished to do. "Once you are educated in the USA," he later explained, "you become accustomed to a very sophisticated approach to engineering. I missed the level of technology, the excitement of responsibility I was given, and the scale of the projects there." He thus returned to Chicago and to SOM in 1960, where he remained for his entire career, rising to the position of general partner and chief structural engineer. Khan was only fifty-two years old when he suffered a heart attack and died in 1982 while visiting Saudi Arabia, where he had designed the Haj Terminal at King Abdul Aziz International Airport at Jidda, an impressive multiple-tent structure that covers 105 acres and can shield from the sun up to eighty-thousand pilgrims at a time on their way to and from Mecca.

Rather than airport terminals or highway bridges, however, it was skyscrapers for which Fazlur Khan came most to be known in engineering and architectural circles worldwide. Indeed, when *Engineering News-Record* bestowed upon him its most prestigious honor by naming him "Construction's Man of the Year" for 1972, the editorial accompanying the cover story spoke of nothing but his tall buildings. At that time, Khan was responsible not only for the world's tallest reinforced-concrete building, Houston's 714-foot-high One Shell Plaza, which had opened the year before, but also for the structural design of the tallest steel building, the 1,450-foot Sears Tower, which was then under construction.

"The social and visual impact of buildings is really my motivation for searching for new structural systems," which made these buildings possible, according to Khan, who had "learned as a structural engineer to think like an architect." His conversance in the two cultures of engineering and architecture was no doubt hastened by his associations at SOM. It was in a question asked by the architect Bruce Graham, with whom Khan became most closely associated, that his principal contribution to the structural design of tall buildings had its origins. The simple question of "what the most economical building would be" elicited from Khan the response that it would be "one with thin solid walls, like a tube, so that it reacted to wind like a vertical cantilever and thus eliminated shear racking," the kind of sideways distortion that a deck of cards displays when pushed from the side.

The tubular principle is common in nature, of course, being present in hollow-stemmed reeds and grasses and in the feather shafts and bones of birds. Khan seems to have been inspired not by nature, however, but by his thinking about how skyscrapers were traditionally framed, with steel or concrete beams and columns arranged in a three-dimensional rectangular grid. He saw that when the wind blew on a very tall building so constructed, each floor was sheared from a true rectangular shape into somewhat of a parallelogram. By locating most of the structural material on the periphery and tying it all together in such a way that the whole structure acted more like a single continuous beam rooted in the ground like a stalk of bamboo, he was able to achieve tremendous stiffness.

Gaining stiffness became essential if taller buildings were to be constructed without paying a premium in structural material. A hundred-story skyscraper, for example, might require twenty-five pounds of steel for every square foot of office space contained in the building, just to hold it up. Such a building would be too flexible, however, if constructed in the way conventional skyscrapers like the Empire State Building were, and the amount of steel would have to be more than doubled to keep the skyscraper from moving too much in the wind. Khan termed the excess weight of steel required for stiffness over strength a "premium for height." Using the tubular principle, however, the structural engineer did not have to pay the premium for a supertall building, and thus Khan "contributed greatly to the renaissance in skyscraper design in the late 60's."

Although the ideal tubular building would have solid structural walls, buildings that are lived and worked in need windows. Khan recognized this, and so he told Graham that "we punch small holes in the tube for windows." Just as the holes we make in floor beams and wall studs to allow wires to pass do not appreciably weaken the structure of a house, so window holes in a tubular frame do not significantly reduce its stiffness. The first application of the tubular principle was in 1963 in Chicago, with the construction of the forty-three-story reinforced-concrete apartment building located at 211 East Chestnut Street.

The first tubular building framed in steel was Chicago's John Hancock Center, completed in 1969. For years before that, the hundred-story skyscraper was news in the engineering and architectural press for the way structural engineers and architects could work together to solve a problem. According to Khan, "the engineer has to be an architect to the extent that the architect has to be an engineer so that in combination they produce the creative building."

The architectural problem posed by the Hancock Center was to provide one million square feet of office space and an equal amount of apartment space, plus about eight hundred thousand square feet for commercial use and parking. Since the center was to be an investment property, the cost of the building was of prime concern. A typical architectural solution would have been to design two buildings, so there would be no premium for height, which the investors did not wish to pay. But two buildings on the single site would have crowded the complex at street level, which the architects did not like. Hence, a structurally efficient single tall building was looked to, and Khan proposed an "optimum column-diagonal truss tube," as he then called the structure that has become so familiar a part of the Chicago skyline. Because the office floors required more expanse than the apartment floors, the building was given its characteristic taper, with the office space located in the wider lower floors. The diagonal bracing, which some feared would make the rental space with obstructed windows less desirable, came to be coveted by tenants as having a cachet that marked it as belonging to the distinctive 1,127-foot-tall building.

To explain why this new building was such an advancement over the skyscrapers of similar height that were built decades earlier, Khan compared the designs of the two eras in a 1967 article in *Civil Engineering*.

His own words are worth quoting extensively to demonstrate how clear thinking, expressed in clear writing, can reduce what might appear to be a terribly complex problem in structural engineering to one readily grasped by engineers and nonengineers alike:

> The answer lies in three basic characteristics of the high-rise buildings built in the thirties. First, a 20-ft column spacing was considered adequate for office spaces. Today a minimum of about 40 ft is considered adequate. In fact, the longer the spacing, the better the office space. Second, the partitions used were generally made of solid masonry from floor to floor, adding considerably to the rigidity of the entire building. Today most partitions are removable, therefore very low in weight and stiffness. Third, the exterior wall detail was generally made of solid masonry or stone, and the window opening consisted of a small percentage of the total wall surface. Today the glass curtain wall is generally attached to the frame as a non-rigid skin.
>
> These three characteristics of the earlier framed structures all added to the lateral rigidity and stiffness of the structural frame. Because of them, a building designed for a sway factor of as much as $\frac{1}{250}$ of its height would probably never sway over $\frac{1}{600}$ under the worst wind loads. It was therefore possible, with many tall buildings built in the thirties, to design the structural frame for strength only and not make extensive checks for the lateral drift under wind loads.

A similar problem presented itself when Sears, Roebuck and Company wanted to build 3.7 million square feet of office space, and the architects "wanted to maintain a decent environment at ground level," which meant providing a plaza that could be "dotted with art pieces." In order to provide the desired floor space, which was to consist of large floor areas in the lower stories, which Sears was to occupy, and decreasing floor areas as the building and the rents rose, Khan proposed a connected cluster of nine tubes, which he often described as "a group of pencils bundled together with a rubber band." Just as such a bundle can stand up by itself much more easily than can a single pencil, so the Sears Tower would be better able to resist the heavy winds to which its record-setting height would be subjected.

Nine tubes, each seventy-five feet square, compose the 225-foot-square Sears Tower, which unlike the John Hancock has parallel sides. To make the building more than a tall and slender rectangular box, the nine tubes are cut off at various heights, giving the building a different appearance when viewed from different angles. This structural/architectural treatment of the tubes also reduces the floor area in stages as the building rises, thus providing a variety of office configurations and sizes. Each of the tubes has a columnless interior, which provides the unobstructed floors that are so desirable in modern office space. At the lower levels, the nine tubes share interior walls of columns, thus economizing on steel and adding additional stiffness to the structure. Khan estimated that without the interior tic-tac-toe grid of columns, the skyscraper would have required twice the thirty-three pounds per square foot of steel that was used. Indeed, had the bundled tubular configuration not been employed in the Sears Tower, he estimated that ten million dollars' worth of additional steel would have been necessary to stiffen the $150 million structure.

By the early 1970s, other engineers had adopted Khan's tubular principle, and it was used in four of the five then-tallest buildings in the world, which included, in addition to Khan's two Chicago skyscrapers, the conventionally framed Empire State Building, the Standard Oil Building in Chicago, and the World Trade Center in New York, the two towers of which were then under construction. More recent applications of a modified tubular principle include the Bank of China Building in Hong Kong and the record-setting Petronas Towers in Kuala Lumpur, Malaysia.

The economic expansion that brought some of the tallest skyscrapers to the Far East might have surprised Fazlur Khan, who like many of his contemporaries and predecessors found opportunities in the United States that did not exist in their native countries at the time. In a story on foreign-born engineers that appeared in *Civil Engineering* in 1980, both Khan and the bridge engineer T. Y. Lin were quoted as saying that they settled in America because it gave them the opportunity to work in a context of high technology and theory rather than relying primarily on experience to design and build structures.

The subjects of the article were also asked to what they attributed the success of foreign-born engineers in America. Khan, no doubt thinking of his own experience, thought it had to do with the fact that many of the more successful ones came from "fairly aggressive, educationally privileged backgrounds," had already endured tough competition in their own countries, and had had the ambition to come to America in the first place. He likened them to "pioneers who opened the American West" and thought that "here their competitive and innovative inclinations flower—they have more freedom to achieve what they want to do." Lin speculated that there might be a high number of successful Asian civil engineers in the United States because "in the developing nations, the more challenging conditions force them to think creatively, to apply new theories to ancient materials and methods." The structural engineer Anton Tedesko, who was educated in Austria, believed that in the years before World War II, at least, civil-engineering education was simply better in Europe than in America, with the exception of a few schools, such as those in Berkeley and Urbana.

The years since the war have seen a change in engineering education in America, however, and a graduate degree from an American institution is widely considered to have become the world standard. Indeed, it is often American-born and American-educated structural engineers who now take leading roles in building some of the most innovative skyscrapers in the Far East. The New York engineer Leslie Robertson is credited with the structural engineering of the twin towers of the World Trade Center and of the Bank of China Building, and Charles Thornton, of the New York firm of Thornton-Tomasetti Engineers, is the engineer credited with the structural concept of the Petronas Towers, which introduced innovative high-strength concrete technology into Malaysia.

But Chicago remains one of the premier cities in the world for structurally significant architecture, and much of that distinction is the legacy of Fazlur Khan. Chicago is also a city noted for its monumental outdoor sculpture, with downtown plazas dominated by works of Picasso, Calder, and Miró. Shortly after Khan's untimely death, the Structural Engineers Association of Illinois initiated an effort to have a sculpture erected to honor his memory, and it commissioned the Spanish sculptor Carlos Marinas to create a bas-relief in stainless steel and bronze. The eleven-foot-long rectangular work has as its focus the head of Khan, flanked by the Chicago skyline and a representation of the structural foundations and frames of some of the buildings. Confusingly, the image of Khan appears to be in a rift between the two sides of the bas-relief, which are offset as if sheared along an earthquake fault, an unlikely event in Chicago. An early rendering of the sculpture shows it without inscription, but when it was unveiled in 1988 it bore superimposed on the cutaway structures the words:

THE STRUCTURAL ENGINEERS ASSOCIATION OF ILLINOIS
RECOGNIZES FAZLUR RAHMAN KHAN AS ONE OF THE GREAT
STRUCTURAL ENGINEERS OF OUR TIME

To the right of Khan's image, over the skyline, he is further identified as "structural engineer, humanist, educator and speaker." How effectively he spoke appears to have made an impression on people, as it did on the reporter who described him as "frequently using his hands like a conductor, sometimes with slide rule in hand, to supplement his precise vocabulary that is devoid of vulgarity, but is spiced with vernacular and some slang."

For all the remarkable qualities of Fazlur Khan that the sculpture evokes, a site for its permanent display was not easily found. The proprietors of the John Hancock Center and Sears Tower did not wish to locate in their lobbies the image of the engineer that made their buildings possible. Neither did the Art Institute of Chicago, the Chicago Academy of Sciences, the Museum of Science and Industry, the Chicago Cultural Center, or the Chicago Historical Society accept the tribute to Khan; nor did the Illinois Institute of Technology, where he had been an adjunct professor of architecture. Though their refusal may have been due in part to what they perceived to be the too prominent place given to the sponsoring organization in the inscription, thus blurring the line between plaque and sculpture, local structural engineers took the rebuff as a further example of the anonymity of engineers, their poor treatment by the media, and their status relative to architects in the Windy City.

On its arrival in Chicago, the work of art was unveiled at an artist's reception in the lobby of an office building and remained on display there for several weeks, after which it was put on display in the lobby of City Hall. Then, with no permanent home, it was put in storage.

Finally, after almost a year, the Brunswick Building, one of Khan's early applications of his tubular-framing principle, agreed to provide a permanent home in its plaza for the sculpture. However, the plaza is dominated by Joan Miró's twenty-five-foot-tall *Miss Chicago*, which dwarfed the four-by-eleven-foot bas-relief mounted on the wall behind it. Making matters worse, after a couple of years of outdoor display, welding on the Marinas sculpture began to rust and deteriorate.

The sculpture finally found an appropriate indoor environment in 1993, when new management at the Sears Tower agreed to mount it in the skydeck area of the building, through which 1.5 million tourists pass each year on their way to the observatory on the 103rd floor, high above Fazlur R. Khan Way.

The Fall of Skyscrapers

The terrorist attacks of September 11, 2001, did more than bring down the World Trade Center towers. The collapse of those New York City megastructures, once the two tallest buildings in the world, signaled the beginning of a new era in the planning, design, construction, and use of skyscrapers everywhere. For the foreseeable future, at least in the western world, supertall buildings would be looked upon as potential terrorist targets, and the continued occupancy of signature skyscrapers by their prestige-seeking tenants would face increased scrutiny.

Since two separate hijacked airplanes loaded with jet fuel were crashed within about fifteen minutes of each other into the two most prominent and symbolic structures of lower Manhattan, the once reassuringly low numbers generated by probabilistic risk assessment seem irrelevant. What happened in New York ceased being a hypothetical, incredible, or ignorable scenario. From then on, structural engineers have had to be prepared to answer harder questions about how skyscrapers will stand up to the impact of jumbo jets and, perhaps more important, how they will fare in the ensuing conflagration. Architects have had to respond more to questions about stairwells and evacuation routes than to ones about facades and spires. Because of the nature of skyscrapers, neither engineers nor architects are likely to be able to find answers that will satisfy everyone.

Although the idea of a skyscraper is modern, the inclination to build

World Trade Center, September 11, 2001

upward is not. The Pyramids, with their broad bases, reached heights unapproached for the next four millennia. But even the great Gothic cathedrals, crafted of bulky stone into an aesthetic of lightness and slenderness, are dwarfed by the steel and reinforced-concrete structures of the twentieth century. It was modern building materials that made the true skyscraper structurally possible, but it was the mechanical device of the elevator that made the skyscraper truly practical. Ironically, it is also the elevator that has had so much to do with limiting the height of most tall buildings to about seventy or eighty stories. Above that, elevator shafts occupy more than 25 percent of the volume of a conventional tall building, and so the economics of renting out reduced floor space argues against investing in greater height.

The World Trade Center towers were 110 stories tall, but even with an elaborate system of local and express elevators, the associated sky lobbies and utilities located in the core still removed almost 30 percent of the towers' floor area from the rentable-space category. By all planning estimates, the World Trade Center towers should have been viewed as a poor investment and so would not likely have been undertaken as a strictly private enterprise. In fact, it was the Port of New York Authority, the bistate governmental entity now known as the Port Authority of New York and New Jersey, that in the 1960s undertook to build the towers. With its toll revenue and ability to issue bonds, the Port Authority could afford to undertake a financially risky project that few private investors would dare.

Sometimes private enterprise does engage in similarly questionable investments, balancing the tangible financial risk with the intangible gain in publicity, with the hope that it will translate ultimately into profit. This was the case with the Empire State Building, completed in 1931 and seventy years later still the seventh tallest building in the world. Although it was not heavily occupied at first, the cachet of the world's tallest building made it a prestigious address and added to its real-estate value. The Sears Tower stands an impressive 110 stories tall, the same count that the World Trade Center towers once claimed. The Chicago skyscraper gained for its owner the prestige of having its corporate name associated with the tallest building in the world. The Sears Tower, completed in 1974, one year after the second World Trade Center tower was finished, held the title for more than twenty years—until the twin Petronas Towers were completed in Kuala Lumpur in 1998, emphasizing the rise of the Far East as the location of new megastructures.

It is not only the innovative use of elevators, marketing, and political will that has enabled supertall buildings to be built. A great deal of the cost of such a structure has to do with the amount of materials it contains, so lightening the structure lowers its cost. Innovative uses of building materials can also give a skyscraper more desirable office space. The steel frame of the Empire State Building has closely spaced columns, which break up the floor space and limit office layouts. In contrast, the World Trade Center employed the tubular-construction principle, in which closely spaced steel columns were located only around the periphery of the building. Sixty-foot-long steel trusses spanned between these columns and the inner structure of the towers, where further columns were located, along with the elevator shafts, stairwells, and other nonexclusive space. Between the core and tube proper, the broad columnless space enabled open, imaginative, and attractive office layouts. Thus, the twin towers in New York were recognized as a remarkable engineering achievement, if not an architectural one.

When new to the Manhattan skyline, the unrelieved 209-foot-square floors and virtually unbroken 1,360-foot-high profiles of the twin World Trade Center buildings came in for considerable architectural criticism for their lack of character. Like the Sears Tower, however, when viewed from different angles, the buildings, especially as they played off against each other, projected a great aesthetic synergy. The view of the towers from the walkway of the Brooklyn Bridge was especially striking, with the stark twin monoliths echoing the twin Gothic arches of the bridge's masonry towers.

Although the World Trade Center towers did look like little more than tall prisms from afar, the play of the ever-changing light on their aluminum-clad columns made them new buildings by the minute. From a closer perspective, the multiplicity of unbroken columns corseting each building also gave it an architectural texture. The close spacing of the columns was dictated in part by the desire to make the structure as nearly a perfect tube as possible. The compromise struck in the World Trade Center was to use tall but narrow windows between the steel columns. In fact, the width of the window openings was said to be less that the width of a person's shoulders, which was intended by the reportedly acrophobic architect, Minoru Yamasaki, to provide a measure of reassurance to the occupants. Since the terrorist attack, however, one of the most haunting images of those windows is of so many people standing sideways in the narrow openings, clinging to the columns and, ultimately falling, jumping, or being carried to their deaths.

Terrorists first attempted to bring the World Trade Center towers down in 1993, when a truck bomb exploded in the lower-level public parking garage, at the base of the north tower. Power was lost in the building, and smoke rose through it. It was speculated that the terrorists were attempting to topple the north tower into the south one, thereby bringing them both down, but even though several floors of the garage were blown out, the structure stood. There was some concern among engineers that the basement columns, no longer braced by the garage floors, would buckle, and so they were fitted with steel bracing before the recovery work proceeded. After that attack, access to the underground garage was severely restricted, and security in the towers was considerably increased. Many people no doubt recalled the 1993 bombing when the airplanes struck the towers in 2001.

As they had the truck bombing, the World Trade Center towers clearly survived the impact of the Boeing 767 airliners. Given the proven robustness of the structures to the earlier assault, the thought that the buildings might actually collapse may have been far from the minds of many of those who were working in them on September 11, 2001. It certainly appears not to have been feared by the police and firefighters who rushed in to save occupants and extinguish the flames. Indeed, the survival of the World Trade Center after the 1993 bombing appears to have given an unwarranted sense of security that the buildings could withstand even the inferno created by the estimated twenty thousand gallons of jet fuel that each plane carried. (That amount of fuel has been estimated to have an energy content equivalent to about 2.4 million sticks of dynamite.)

Steel buildings are expected to be fireproofed, and so the World Trade Center towers were. However, fireproofing is a misnomer, for it only insulates the steel from the heat of the fire for a limited period, which is nevertheless supposed to be enough time to allow for the fire to be brought under control, if not extinguished entirely. Unfortunately, conventional firefighting means, such as water, have little effect on burning jet fuel. In fact, the burning fuel is believed to have been consumed rather quickly, but the office furnishings and products that it ignited continued to burn, and water to fight that fire was unavailable after the crash. Thus, the fire continued unabated. It has been speculated that some of the steel beams and columns of the structure may have been heated close to if not beyond their melting point.

Even if the steel did not melt, the effects of prolonged elevated temperatures caused it to expand, soften, sag, bend, and creep. The intense heat also affected the concrete floor, which, no longer adequately supported by the structure in place before the impact, began to crack, spall, and break up, compromising the synergistic action of the parts of the structure. Without the stabilizing effect of the stiff floors, the steel columns still intact became less and less able to sustain the load of the building above them. When the weight of the portion of the building above became too much for the locally damaged and softened structure to withstand, it collapsed onto the floors below. The impact of the falling top of the building on the lower floors, the steel columns of which were also softened somewhat by heat conduction along them, caused them to collapse in turn, creating an unstoppable chain reaction, termed "pancaking." The tower that was struck second failed first in part because the plane hit lower, leaving a greater weight to be supported above the damaged area. (The collapse of the lower floors of the towers under the falling weight of the upper floors occurred for the same reason that a stack of books easily supported on a coffee table can break that same table if dropped on it from a sufficient height.)

Within days of the collapse of the towers, failure analyses appeared on the Internet and in engineering classrooms. Perhaps the most widely circulated were the mechanics-based analysis of Zdenek Bazant of Northwestern University and the energy approach of Thomas Mackin of the University of Illinois at Urbana-Champaign. Each of these estimated that the falling upper structure of a World Trade Center tower exerted on the lower structure a force some thirty times what it had once supported. Charles Clifton, a New Zealand structural engineer, argued that the fire was not the principal cause of the collapse. He believed that it was the damaged core rather than the exterior tube columns that succumbed first to the enormous load from above. Once the core support was lost on the impacted floors, there was no stopping the progressive collapse, which was largely channeled by the structural tube to occur in a vertical direction.

However the collapse occurred, in the wake of the disaster the immediate concerns were, of course, to rescue from the rubble as many people as might have survived. Unfortunately, even to recover most of the bodies proved an ultimately futile effort. The twin towers were enormous structures. Each floor of each building encompassed an acre, and the towers enclosed sixty million cubic feet each. Together, they contained 200,000 tons of steel and 425,000 cubic yards (about 25,000 tons) of concrete. The pile of debris in some places reached as high as a ten-story building. One month after the terrorist attack, it was estimated that only 15 percent of the debris had been removed, and it was thought that it would take a year to clear the site. However, with respect, dedication, and resolve, the army of workers at Ground Zero, as the site came to be called, cleared it out to bedrock months sooner.

Among the concerns engineers had about the cleanup operation was how the removal of debris might affect the stability of the ground around the site. Because some of the land on which the World Trade Center was built had originally been part of the Hudson River, an innovative barrier had been developed at the time of construction to prevent river water from flowing into the basement of the structures. This was done with the construction of a slurry wall, which began as a deep trench that was filled with a mudlike mixture until a hardened reinforced-concrete barrier was in place. The completed structure provided a watertight enclosure, which came to be known as the "bathtub," within which the World Trade Center towers were built. The basement floors of the twin towers acted to stabilize the bathtub walls, but these floors were crushed when the towers broke up and collapsed into the enclosure. Early indications were that the bathtub remained intact, but in order to be sure its walls did not collapse when the last of the debris and thus all internal support was removed, vulnerable sections of the concrete wall were tied back to the bedrock surrounding the site even as the debris removal was proceeding.

Atop the pile of debris, the steel beams and columns were the largest and most recognizable parts in the wreckage. The concrete, sheetrock, and fireproofing that were in the building were largely pulverized by the collapsing structure, as evidenced by the ubiquitous dust present in the aftermath. (A significant amount of asbestos was used only in the lower floors of one of the towers, bad publicity about the material having accelerated during the construction of the World Trade Center. Nevertheless, in the days after the collapse, the once-intolerant Environmental Protection Agency declared the air safe.) The grillelike remains of the buildings' facades, towering precariously over Ground Zero, became a most eerie image. Though many argued for leaving these cathedral wall-like skeletons standing as memorials to the dead, they posed a hazard to rescue workers and were in time torn down and carted away for possible future reuse in a reconstructed memorial. As is often the case following such a tragedy, there was also some disagreement about how to treat the wreckage generally. Early on, there was clearly a need to remove as much of it from the site as quickly as possible so that any survivors there could be uncovered. This necessitated cutting up steel columns into sections that could fit on large flatbed trucks. Even the relocation of so much wreckage presented a problem. Much of the steel was marked for immediate recycling, but forensic engineers worried that valuable clues to exactly how the structures collapsed would be lost.

All of the speculations of engineers about the mechanism of the collapse were and are in fact hypotheses, theories of what might have happened. While computer models can be constructed to test hypotheses and theories, actual pieces of the wreckage may provide the most convincing confirmation that the collapse of the structures did in fact progress as hypothesized. Though the wreckage may appear to have been hopelessly jumbled and crushed, telltale clues could have survived among the debris. Pieces of partially melted steel, for example, could provide the means for establishing how hot the fire burned and where the collapse might have initiated. Badly bent columns can give evidence of buckling before and during collapse. Even the scratches, gouges, and scars on large pieces of steel can be useful in determining the sequence of collapse. This was to be the task of teams of experts announced shortly after the tragedy by the American Society of Civil Engineers and other professional societies in conjunction with the Federal Emergency Management Agency. Also in the immediate wake of the collapse, the National Science Foundation awarded grants to engineering and social-science researchers to assess the debris as it was being removed and to study the behavior of emergency-response and emergency-management teams.

Analyzing the failure of the towers presents a herculean task, but it is important that engineers understand in detail what happened so that they can incorporate the lessons learned in future design practices. It was the careful failure analysis of the bombed Murrah Federal Building in Oklahoma City that led engineers to delineate guidelines for designing more terrorist-resistant buildings. The Pentagon was actually undergoing retrofitting to make it better able to withstand an explosion when it was hit by a third hijacked plane on September 11. Part of the section of the building that was struck had in fact just been strengthened, and it suffered much less damage than the old section beside it, thus demonstrating the effectiveness of the work.

Even before a detailed failure analysis was completed, however, it was evident that one way to minimize the damage to tall structures is to prevent airplanes and their fuel from being able to penetrate deeply into the buildings in the first place. This is not an impossible task.

When a B-25 bomber struck the Empire State Building in 1945, its body stuck out from the seventy-eighth and seventy-ninth floors like a long car from a short garage. The building suffered an eighteen-by-twenty-foot hole in its face, but there was no conflagration, and there certainly was no collapse. The greatest damage was done by the aircraft's engines coming loose and flying like missiles through the building. The wreckage of the airplane was removed, the local damage repaired, and the building restored to its original state. Among the differences between the Empire State Building and World Trade Center incidents is that in the former case, relatively speaking, a lighter plane struck a heavier structure. Furthermore, the propeller-driven bomber was on a short-range flight, from Bedford, Massachusetts, to Newark, New Jersey, and so did not have on board the amount of fuel necessary to complete a transcontinental flight or sufficient to bring down a skyscraper.

Modern tall buildings can be strengthened to be more resistant to full penetration by even the heaviest of aircraft. This can be done by placing more and heavier columns around the periphery of the structure, making the tube denser and thicker, as it were. The ultimate defense would be to make the facade a solid wall of steel or concrete or both. This would eliminate windows entirely, of course, which would defeat some of the purpose of a skyscraper, which is to provide a dramatic view from a prestigious office or boardroom. The elimination of that attraction, in conjunction with the increased mass of the structure itself, would provide space that would cost a great deal more to build and yet command a significantly lower rent. Indeed, even those who prize safety above all else would be unlikely to consider building or renting space in such a structure. The solution would be a Pyrrhic victory over terrorists.

The World Trade Center towers likely would have stood after the terrorist attack if the fires had been extinguished quickly. But even if the conventional sprinkler systems had not been damaged, water would not have been very effective against burning jet fuel. Perhaps skyscrapers could be fitted with robust firefighting systems employing the kind of foam that is laid down on airport runways during emergency landings or fitted with some other oxygen-depriving scheme, if there could also be a way for fleeing people to breathe in such an environment. Such systems would need to possess a robustness and redundancy to survive tremendous impact forces, and they might be unattractively bulky and prohibitively expensive to install. Other approaches might include more effective fireproofing, such as employing ceramic-based materials, if they could be made shatterproof, thus at least giving the occupants of a burning building more time to evacuate.

The evacuation of tall buildings is now given much more attention by architects and engineers alike. Each World Trade Center tower had multiple stairways, but all were in the single central core of the building. In contrast, stairways in Germany, for example, are required to be in different corners of the building. In that configuration, it is much more likely that at least one stairway will remain open if, for example, an airplane crashes into another corner. But locating stairwells in the corners of a building means, of course, that prime office space cannot be located there. In other words, most measures to make buildings safer also make them more expensive to build and diminish the appeal of their space. This dilemma is at the heart of the reason why the future of the supertall building is threatened.

It was unlikely that, in the immediate wake of the World Trade Center collapses, any supertall building then in the development stage, in the United States at least, would not have been put on hold and reconsidered. Real-estate investors want to know how a proposed building will stand up to the crash of a fully fueled jumbo jet, how hot the ensuing fire will burn, how long it will take to be extinguished, and how long the building will stand so that the occupants can evacuate. They also raise the question of who will rent the space if it is built.

Potential tenants naturally have the same questions about terrorist attacks. Companies wonder if their employees will be willing to work on the upper stories of a tall building. Managers wonder if those employees who do agree to work in the building will be constantly distracted, watching out the window for approaching airplanes. Corporations wonder if clients will be reluctant to come to a place of business perceived to be vulnerable to attack. The very need to have workers grouped together on adjacent floors in tall buildings is called into question.

After the events of September 11, the incentive to build a signature structure, a distinctive supertall building that sticks out in the skyline, was greatly diminished. In the wake of the tragedy, as leases came up for renewal in existing skyscrapers, real-estate investors watched closely the downward movement in occupancy rates. It may be that some of our most familiar skylines may be greatly changed in the foreseeable future. Indeed, if companies move their operations wholesale out of the most distinctive and iconic of supertall buildings and into more nondescript structures of moderate height, it is not unimaginable that cities such as New York and Chicago will in time see the reversal of a long-standing trend. We might expect no longer to see developers buying up land, demolishing the low-rise buildings on it, and putting up a taller sky-scraper than those in the vicinity. Instead, owners might be more likely to disassemble the top of or entirely demolish a vacant skyscraper and erect in its place a building that is not significantly smaller or taller than its neighbors. Skylines that were once immediately recognizable—even in silhouette—for their peaks and valleys may someday be as flat as a mesa.

There is no imperative to such an interplay between technology and society. What happens ultimately will depend largely on how governments, businesses, and individuals react to terrorism and the threat of terrorism. Unfortunately, the image of the World Trade Center towers collapsing will remain in our collective consciousness for a few generations, at least. Thus, it is no idle speculation to think that it will be at least a generation before skyscrapers return to ascendancy, if they ever do, at least in the west. Developments in microminiaturization, tele-communications, information technology, business practices, management science, economics, psychology, and politics will likely play a much larger role than architecture and engineering in determining the immediate future of supertall buildings and other macrostructures.

Vanities of the Bonfire

History has been punctuated regularly by colossal structural failures. The final configuration of the bent pyramid, completed almost four millennia ago in Dahshur, is believed to have resulted from its initially being built at the overly ambitious angle of fifty-four degrees. After a landslide of stone during construction, the builders apparently lowered their sights and changed the top section to a forty-three-degree incline. The thirteenth-century collapse of the cathedral at Beauvais marked the end of an era in Gothic building during which "taller" and "lighter" had been the watchwords. In more modern times, the tendency to build ever longer and more slender bridges led to such catastrophic failures as the collapse of the Quebec cantilever bridge during construction and of the infamous Tacoma Narrows suspension bridge just four months after it was completed. Such tragedies are rooted in two human characteristics: the cultural drive to build ever-bolder structures and the hubris of master builders and engineers in their attempts to do so.

A recent example of the tragic failure of a construction project fully embodied these all-too-human characteristics. On November 18, 1999, the massive pile of logs known as Bonfire collapsed spontaneously at Texas A&M University, taking twelve lives and injuring dozens of other students. In the wake of the tragedy, the university president appointed a special commission to investigate the causes of the accident. Its report was issued within six months and provides an insightful look into not only the mechanical causes of such an accident but also the behavioral causes stemming from the interaction—or, rather, the lack thereof—between the student designers and builders of the structure and the university administration. Such a context often comes under the rubric of the prevailing culture, which has to be experienced to be fully appreciated.

About ten years before the fatal bonfire, I visited Texas A&M the week before Thanksgiving to deliver a lecture in the civil engineering department. After the lecture, my host told me there were two things I had to experience before leaving town. One was a famous local barbecue establishment, in which the brisket was served with whole cooked onions, a pickle, a wedge of cheddar cheese, and some bread—all wrapped up in a piece of butcher paper that when opened served as place mat and plate. The only utensil provided was a very sharply pointed knife, which was used by my host to cut, spear, and convey the beef to his mouth. It made me nervous to see him wield the knife so casually, all the while joking that he hoped no one would come up behind him and slap him on the back just as he was putting the knife in his mouth. Living on the edge was presented to me as part of the Aggie dining experience.

The second thing my host made sure that I saw before leaving town was the bonfire stack, which was being readied in anticipation of the big football game between A&M and archrival University of Texas at Austin. The concept of a bonfire was not new to me, for the students at my own university were fond of burning fraternity benches to celebrate a victory over their basketball rival from Chapel Hill. On some occasions, these bonfires had gotten a bit out of hand, with the crowd sometimes pushing those in the inner circle too close to the fire for comfort. A particularly dangerous practice involved students, their judgment and coordination impaired by celebratory spirits, jumping through the fire as if they were celebrating some pagan rite. Some of these students were badly burned when they happened to trip on the burning debris and fall onto the fire. In response to such incidents, the university has alternated between totally banning bonfires and allowing them to proceed only under controlled circumstances.

At Texas A&M, the bonfire has been raised to new heights, however.

Rivalries are intense and tradition-laden in Texas, and the one between Texas A&M University and the University of Texas is perhaps the most intense, complex, and tradition-laden of all. The annual football game between the Aggies and the Longhorns takes place on or around Thanksgiving, the venue naturally alternating between the two campuses. And since 1909 the football game had been preceded in College Station by a bonfire.

The first bonfire consisted of a pile of trash set ablaze and was most likely an ad hoc affair. In 1912 lumber intended for the construction of some dormitories was surreptitiously diverted to the bonfire. After 1935, when a farmer's log barn was "acquired" and burned, the bonfire tradition was regulated by the college. The first all-log bonfire was constructed in 1943. At twenty-five feet tall, it was a modest structure by later standards. By midcentury, a spliced center pole was being used to erect stacks of logs more than fifty feet tall. The height of the bonfire continued to grow, reaching almost 110 feet in 1969. It was at this point that the university, which had grown in stature along with its bonfire, restricted the size of the structure to fifty-five feet tall and forty-five feet in diameter. At some point, the tradition, the process, and the stack of logs itself came to be referred to—without an article, but capitalized—simply as Bonfire.

The manpower—until 1979 women were not involved in Bonfire—needed to bring the cut logs from a site in want of clearing to a towering stack to be ignited on the eve of the big game has been estimated at 125,000 man-hours distributed among about five thousand workers. It has also been conjectured that as many as eight thousand logs have been brought to a bonfire site, there to be wired together into a multitiered wedding cake–like structure built around the center pole. The entire process usually began in early October, with the stacking of the logs alone taking as many as three weeks. The structure was topped by an outhouse, referred to as the t.u. (as the Aggies refer to the University of Texas, known to Longhorns as U.T.), tearoom, or frat house. When the outhouse-topped log stack was complete and ready to be set afire, it was doused with about seven hundred gallons of diesel fuel. This icing on the cake, as it were, was traditionally applied by the Texas Engineering Extension Service Fire Training School.

To many an outsider, in retrospect as in prospect, the size of the

Aggie bonfire seemed to be a great excess. The practice was clearly fraught with danger, like eating Texas barbecue with a pointed knife. But tradition knows little fear and shows little respect for the opposition, whether it be another university's football team or the force of gravity.

Bonfire over the years generally proceeded on schedule and according to plan. There were attempts by rival fans to ignite a bonfire early, including the dropping of firebombs from an airplane and the planting of explosives, but none had been successful. Until 1999, all but one Bonfire was lit by the Aggies as scheduled; the 1963 Bonfire was not ignited at all because of the assassination of President Kennedy. And in 1999, Bonfire was not ignited because the structure collapsed while still under construction, on the Thursday before Thanksgiving. By that Sunday, Texas A&M president Ray M. Bowen had announced the establishment of a commission "charged to initiate a review of all aspects of the 1999 Aggie bonfire and to examine evidence developed by other investigations" of the tragedy. The special commission was chaired by Leo Linbeck, Jr., a Houston construction executive.

Within six months, the final report of the special commission was issued, and it revealed and acknowledged officially design factors and patterns of behavior on the part of those participating in Bonfire that contributed to the accident. The report shows the structural collapse to be a classic case of design evolution and engineering hubris contributing to what in retrospect appears to have been an accident waiting to happen. Bonfire tradition was to build on the successes of past years, but modifications made from year to year negated what could be learned from experience. The builders took the energy stored in a two-million-pound stack of logs a little too lightly, and they approached the construction problem as if it were actually a piece of cake.

Even before the commission had begun its methodical study of the catastrophic failure, there were theories about the collapse, as there are with any structural accident with so many fatalities and such visibility. Among the prime early suspects was the tall center pole, which was made by splicing two standard utility poles together along an elaborately fashioned lap joint. A Bonfire proper may be said to have started with the raising of the center pole, which was buried as many as fifteen feet in the ground and steadied by guylines anchored to other, outwardly inclined poles spaced around the perimeter of the construction site. The use of a center pole was introduced in the mid-1940s, when the configuration of the bonfire stack was conical, like a tepee, a shape achieved simply by leaning logs against the center pole. The height of such an arrangement was limited by the length of the logs used, but in the late 1950s a tepee Bonfire could reach a height of seventy feet. Such a height required finding sufficiently long logs to lean against the pile, and it was their scarcity that led in the 1960s to the development of the wedding-cake style of log stack.

The center pole serves not only as a symbolic axis for Bonfire but also enables the use of block and tackle to assist in raising logs and workers as the stack rises. The 1999 Bonfire stack had reached about forty-five feet up the center pole when the accident occurred. Students who witnessed the collapse reported that they noticed the stack begin to shift and then heard a loud crack, followed by the collapse of the incomplete structure. Some observers interpreted this sequence of events as pointing to the fracture of the center pole as the initiator of the fatal event. However, after a structural analysis, the commission found that "given the enormous weight of the stack, even a perfect center pole could not have played a significant role in providing structural strength." In other words, for all its symbolic function, the center pole had never supported the stacks of logs piled around it, and so another cause had to be found.

The soil on which Bonfire was built had also been an early suspect in the collapse. This seemed to provide a credible explanation, given the fact that in 1994 soil softened by rain was identified as the reason that the pile of logs fell over just two weeks before the big football game. However, the fall of 1999 was not as wet as that of 1994, and, as reported by the special commission, "analysis showed the soil to be sufficiently compact and stable and that it could easily support a structure at least twice as heavy."

The guys steadying the center pole were also looked to as possible causes of the collapse, but "all ropes tested were of good quality." Furthermore, "although one of the guy ropes did fail during the collapse, it was not a contributing factor because it broke after the collapse sequence had started."

Texas A&M Bonfire stack, 1994

An incident that occurred a few days before the collapse came under suspicion as well. Witnesses reported that a crane used to lift logs into place had hit one of the cross ties fastened to the center pole, breaking off a piece of the tie. However, the commission found that the force that would have accompanied such an impact "could not have materially affected the center pole or contributed to the collapse."

Even a strong wind, an earthquake, ground movement associated with trains passing nearby, and sabotage were looked to as possible initiators of the accident, but the commission's analysis could give no credence to such causes. Furthermore, no defects were found in any of the perimeter poles used to anchor the guys or in any structural member or piece of equipment used in the construction. In other words, the causes of the collapse were in fact subtle and, according to the commission,

> [t]he engineering analysis of the Bonfire collapse turned out to be much more challenging than originally anticipated. The physical factors ultimately determined to be drivers of the collapse were not obvious to the engineering teams at the outset. In fact, it took a number of weeks and considerable effort before the collapse mechanism and sequence were determined.

Those efforts included the development of a composite design of "the historical Bonfire" and its examination by means of the general-purpose finite-element-analysis computer program ABAQUS. The computer-based model enabled an engineering team to simulate hypothesized conditions in the ill-fated log stack and to confirm its posited behavior. In the final analysis, the commission found the collapse to be driven by a combination of factors rather than any single factor, and each of those factors pointed to a mind-set among the university's students and administration characterized by complacency, hubris, and a disrespect for the forces of nature:

1. *The bonfire was built on slightly sloping ground.* Although the ground was solid, it was not level, dropping about one foot from the northwest to the southeast side of the structure, which was on the order of fifty feet across. This meant that the first tier of logs leaned to the southeast. The upper tiers of logs and the tall center pole they were built around were aligned with the true vertical, however, creating a bent structure not unlike the Tower of Pisa. For a two-million-pound tower of logs to be built in this manner is to invite instability, and the structure did in fact collapse downhill to the southeast.

2. *The logs used were more crooked than usual.* In previous bonfires, the logs used were very straight and so fit closely together, like uncooked spaghetti held tightly in the fist. The logs used in the fatal stack, by being

more crooked than usual, allowed numerous gaps to exist among the logs in the lower tiers. This feature might actually have been seen as a plus by the bonfire erectors, since upper-tier logs could be inserted into the gaps, thus providing an interconnection between tiers. In fact, rather than providing a beneficial interconnection, the logs so used proved to be a major contributor to the collapse.

3. *Upper-tier logs were wedged between lower-tier logs.* The advantage of interconnection became a disadvantage when the second-tier logs were wedged so tightly and so deeply into the tier below that additional outward pressure was created in the foundation stack. Because wedging was used more aggressively in 1999 than in previous bonfires, the lower stack was like an already full pencil holder being stuffed with more and more pencils. In effect, the stack was filled to bursting.

4. *The upper tiers of logs were built out farther than in past years.* After Bonfire reached 109 feet high, in 1969, there were restrictions imposed on the height and width of the stack of logs. However, the width restriction of forty-five feet was interpreted to apply only to the base of the stack and to place no restrictions on higher levels. In order that Bonfire contain as many logs as possible, the 1999 structure was being constructed with wider upper stacks. Like a skyscraper built without regard for setback restrictions, the bonfire had a larger than anticipated volume and therefore bore down with a greater weight on its lower levels. This additional weight caused the wedged logs to be driven even deeper into the tiers below and created still further outward pressure on the ground-level logs in the bonfire.

5. *Steel cables were not wrapped around the lowest logs.* In previous bonfires, the lowest tier of logs was held together by steel cables wrapped around the outside of the entire bundle. However, there had been some disappointment in recent years that bonfires were burning too quickly, and this was attributed by some students to the use of the cables. For this and other reasons, steel cables were not used in the fatal stack, perhaps in part because it was thought, incorrectly, that the effects of some of the other modifications, like wedging, were mitigating. In effect, the ground-level stack of logs was constructed like a barrel without barrel hoops, leaving the staves free to expand under the pressure of the barrel's contents. The collapse does indeed appear to have been triggered by

this bursting of the bottom tier, with the falling stack bringing the center pole and the unfortunate students down with it.

As important as it is to understand the role that physical factors play in the collapse of a structure of the magnitude of Bonfire, there are human considerations that must also be understood to grasp fully why decisions that in retrospect so adversely affected safety were made in the first place. The special commission cited these behavioral factors and considered them in detail, using a "behavioral cause analysis" that relied heavily on the idea of identifying what are metaphorically described as "holes in barriers." According to the team responsible for the behavioral analysis:

> Today, few catastrophic events are caused by simple human error. Modern systems have defense-in-depth in the form of multiple barriers to prevent events. Examples of these barriers include training, procedures, inspections, and reviews. So now when a significant event is experienced, there is virtual certainty that several causes acted together to both trigger the event sequence and to fail all of the barriers provided to prevent the event.

As a result of their use of "barrier analysis as a method of cause analysis," the behavioral-science team identified four "root causes" of the collapse:

1. *Bonfire was designed without adequate engineering analysis.* Decisions regarding the design of Bonfire were made by the Red Pots, a group of nine juniors and nine seniors so named because of the color of the helmets they wear. The Red Pots, although not experienced structural engineers, were allowed to make crucial decisions regarding size, wedging practices, steel-cable use, and the like for a structure of major proportions. As long as Bonfire took the form of a relatively simple and stable tepee design, the decisions of the Red Pots were not so crucial. However, with the introduction of the wedding-cake configuration, the structural behavior of Bonfire became more complex and nonintuitive. The conventional wisdom among Bonfire enthusiasts was that the structure was safe because it had worked successfully for many years, even though it was being modified from year to year.

2. *Crucial details of Bonfire design were not documented.* The accident investigators could find no evidence that "critical design attributes" that insured adequate safety margins were documented in drawings, specifications, or procedures. Although the Red Pots appeared to have allowed only what were considered small and insignificant changes from a "historically proven design," in fact the cumulative effect of these changes led to a design that was not historically proven.

3. *The university did not acknowledge the magnitude of the danger.* The behavioral scientists call the "organizational equivalent of tunnel vision" cultural bias, and the bias of the Texas A&M culture was to not recognize that "the Bonfire structure had grown too large to be constructed using past practices." Even though A&M is an institution known for its technical prowess, engineering and construction-sciences faculty members were not especially involved with Bonfire. Those few specialist faculty members who were involved "focused on improving the structure to extend the time the fire burned before collapse of the structure. They did not interpret the performance problems as symptoms of structural instability."

As further evidence of the institution's hands-off approach to Bonfire, the structure's height routinely exceeded the administration's fifty-five-foot height limit, which was imposed to reduce the risk of the fire spreading to nearby buildings. When concerns continued to arise about this danger, rather than enforcing strict size limitations on Bonfire, it was moved to the Polo Field, the site of the fatal accident.

Though various concerns over Bonfire occurred over the years, the question of structural design or stability never appears to have been studied by the administration or anyone else. According to the Texas A&M University special commission, "No credible person ever suggested to TAMU administration that the Bonfire structure was unsafe. However, evidence suggests that TAMU administration and staff should have recognized several precursor events as indications that the structure had small safety margins." Among these events were the structural collapses in 1957 and in 1994 and the "steady decrease in time of burn before collapse to an approximate mean of 30 minutes for years 1995–1998." In response to the last indication, structural changes were made to increase burn time, apparently without regard to "structural integrity during construction." Through a naïve logic incomprehensible to a structural engineer, it was to increase burn time that the steel cables were left off the lower stacks of the 1999 Bonfire. According to one student speaking of Bonfire, "From the outside you can't understand it, and from the inside you can't explain it."

4. *Student organizations did not heed warnings that Bonfire was unsafe.* There was clear evidence that Bonfire was not supervised by the students themselves as well as should have been expected. Among the evidence were the facts that injury rates were several times what they are in the construction and forestry industries; that injury rates demonstrated a steadily increasing trend; that fatalities had occurred; that falls from the Bonfire stack occurred repeatedly; that hazing and harassment took place, even though forbidden; and that there were alcohol-related incidents that compromised safety.

In the end, the behavioral-failure analysis found no specific individuals to blame for the fatal accident. According to the report,

> [t]he 1999 Bonfire Structure Collapse is neither a 1999 problem nor a 1999 Red Pot problem. The 1999 Bonfire Structure Collapse is a classic example of an organizational accident with failure causes that existed for many years before the event. No one person in Bonfire performed at such a substandard level as to directly cause the collapse. . . .
>
> For modern era TAMU administrations, Bonfire was and is an institution. Leaders generally do not change institutions unless there is a perceived need for change, and in this case no one noticed the mounting risk.
>
> Bonfire grew. Bonfire grew in size (from a trash fire to a structure), grew in complexity (from a single tiered cone to a multi-tiered wedding cake), grew in the number of people . . . and grew in the number of problems. Most relevant to the 1999 collapse, the structure grew from a simple one that could be "designed" and constructed by students to a complex and risk-significant one that could not. Red Pots continued to maintain the design of a complex structure through an oral tradition. As a result, Bonfire was never built the same way twice even though the accepted basis for safe design was "we have always done it this way and it always worked."

Ironically, the strong tradition of Bonfire at Texas A&M, which could have been the source of a long institutional memory about the dangers and pitfalls associated with such a major structural undertaking, was in fact an impediment to safe practice. Rather than encouraging the sharing of an institutional memory that might have prevented the fatal accident, the tradition had developed into a separation of the mind from the body. What had evolved into such a strong tradition on the surface had, in fact, devolved into an almost traditionless and ad hoc practice when it came to the crucial structural details of Bonfire. It was as if an engineering office had maintained the appearance of order by presenting its plans in carefully color-coordinated reports with eye-catching logos but had given its engineering interns free rein to devise their own plans and carry out their own calculations, all of which were accepted without being challenged or inspected.

In reading the report of the special commission, one sees a striking divide between the students and the administration—between what should have been the junior and senior partners in the undertaking. Apparently the long-standing success of Bonfire, in which the overwhelming majority of the extended Texas A&M community took great pride, had made all the participants overconfident and suppressed the legitimate concerns of the few naysayers.

Seven months after the accident and six weeks after the release of the investigative report, President Bowen announced that Bonfire would be suspended for at least two years. This hiatus would allow time for a new university committee to work on reorganizing Bonfire and for planning to proceed for a scaled-down and tightly supervised event, which could resume as early as the fall of 2002. Among the changes instituted by the Texas A&M administration was the length of time devoted to Bonfire. Students would no longer cut down the trees used—an activity that had taken as long as three months—thus reducing the time they would be allowed to spend on Bonfire to about two weeks. The design of the tower would revert to the tepee shape, and its size would be reduced and enforced, thus also addressing the concerns that had been growing over the environmental waste represented in burning so many trees.

Finally, professional engineers would work on the design of the log structure and would prescribe its construction sequence. This anticipated any rulings of the Texas Board of Professional Engineers, which continued to investigate the implications of the 1999 failure. (The Texas Engineering Practice Act requires that licensed professional engineers be involved in the design of and supervise the construction of major structures involving public health, welfare, or safety. The Board of Professional Engineers continued to look into the implications of the act for Bonfire, with the interpretation of the governing law appearing to revolve around the question of whether Bonfire is a "public work.")

Early in 2002, the year Bonfire was to resume, President Bowen announced that the resumption of the tradition would be delayed at least another year. Among the reasons he gave were safety, liability, and cost. One safety consultant who had been engaged withdrew when his firm could not obtain insurance. In spite of the university's efforts to engage another professional safety consultant, no qualified expert submitted a bid for the job, evidently out of concerns over liability. Indeed, according to Bowen, all those who might have had a responsible role in Bonfire 2002 inquired about protection from legal liability. The only way to achieve this was through enormously expensive insurance, and this cost coupled with that of building a redesigned Bonfire—estimated to be in the range of $1 million to $1.5 million—was criticized as excessive, given budget problems faced by the institution. In conjunction with the safety and liability issues, the cost was judged prohibitive. Anticipating questions about resuming Bonfire in 2003, Bowen noted that he would be stepping down as president well before that decision would have to be made. He stated that he was not doing anything to "take away options for the future." Supporters of Bonfire found venues off campus in attempts to continue the tradition but with no assurance of safety.

If the Texas A&M Bonfire had not collapsed in 1999 and instead had been allowed to continue in the laissez-faire manner of the 1990s, some future bonfire likely would have led to a tragedy that demanded a reassessment of the practice. The virtually unregulated evolution of the design of such a massive structure was a prescription for disaster. It is human nature to build on past successes with a bravado, of which students especially have a great deal, that so often can be checked only by tragedy. Had anyone pointed out before the fact the dangers of the individual acts of abandon identified afterward, they no doubt would have been scoffed at, for Bonfire had been such a successful and revered tradition. Unfortunately, it was a tradition carried forward without conservatism. In that regard, the 1999 Bonfire collapse repeated the pattern of a great number of other colossal failures that have plagued amateur and professional builders alike throughout history.

St. Francis Dam

Los Angeles would not have grown into the metropolis that it is today were it not for the expansion of its water supply. In 1900, the city's population was about one hundred thousand and growing rapidly, reaching 175,000 within five years. Since the Los Angeles River watershed was capable of supporting only about two hundred thousand people, the city had the choice of limiting growth or finding new sources of water. A drought in 1904 raised the issue to crisis proportions.

Los Angeles's need to import water had been foreseen a decade earlier by Fred Eaton, who as city engineer had looked for alternative sources of water in the Sierra Nevadas and had identified the Owens Valley, north of the city, as a likely candidate. In the meantime, the U.S. Reclamation Service had begun looking into the feasibility of designing an irrigation scheme for the farmers of the Owens Valley, and Los Angeles had to act fast if it was to obtain the water rights. Eaton took William Mulholland, manager of the newly formed Los Angeles Bureau of Water Works and Supply, to the valley to investigate the possibility of constructing a gravity-flow aqueduct from there to the city nearly 250 miles south. The distance was unprecedented. The longest Roman aqueducts were less than sixty miles long, and New York's Croton Aqueduct was even shorter. However, Owens Lake was more than three thousand feet higher than the city, providing a much greater average gradient than existed in the successful Croton Aqueduct. Thus, the engineering problems, which would involve inverted siphons and pressure tunnels to get the water over and through the mountains in the way, seemed solvable.

Mulholland was an engineer of the old school, which essentially means that he had learned by doing. He was born in Ireland in 1855, went to sea at fifteen, landed in New York City four years later, worked at a variety of jobs in the East and Midwest and then sailed via the Isthmus of Panama to San Francisco. He settled in the Los Angeles area at the age of twenty-two, working as a "water ditch tender" with the Los Angeles City Water Company, a small private provider. According to the retrospective account by geological engineer J. David Rogers,

> Mulholland later recalled that he became interested in things technical when serving as a helper on a drill rig digging water wells that pierced a buried tree trunk at a depth of 600 feet. He went to the library to investigate the manner by which a tree could become buried at such great depth, and read University of California Professor John LeConte's *Introduction to Physical Geology*. Mulholland liked the subject matter so much that he later recalled, "Right there I decided to become an engineer." . . . In the apprenticeship tradition of that era, he sought an increasingly technical workload and educated himself through reading.

Mulholland eventually became general manager and chief engineer of the water company. He proved himself to be so knowledgeable about the poorly documented infrastructure and workings of the water-distribution system that, when the company was acquired by the city in 1902, the self-taught engineer was retained as its manager. It was in this capacity that he accompanied Fred Eaton to the Owens Valley and secured $1.5 million from the Los Angeles Board of Water Commissioners for engineering studies of the situation.

The scale of the project and Mulholland's "lack of substantive experience in constructing such facilities" were used by "other engineers, newspaper editors and electric power interests" to discredit the scheme. In response to the criticism, the city commissioners appointed an Aqueduct Advisory Board, comprising three distinguished consulting engineers, to "make an independent evaluation of the proposed aqueduct design." One of the consultants was John R. Freeman, who had been among the principal designers of the New Croton Aqueduct and who had served on the advisory board to review the design of the Panama Canal. When the external board found Mulholland's aqueduct design "admirable in conception and outline," criticism was quelled. The twenty-three-million-dollar bond issue passed overwhelmingly in 1907.

Construction of the 233-mile aqueduct and its initial filling was not without incident. Upon first carrying water, one of the major siphons in the aqueduct began leaking and was lifted up by the resulting hydraulic forces. The seepage from the riveted steel conduit also triggered a landslide in which the pipe became entangled. Such setbacks were forgotten by most of the thirty thousand people who gathered on November 5, 1913, to watch the opening of the world's longest aqueduct, which was capable of transporting 258 million gallons per day to Los Angeles.

While the aqueduct was being planned, speculators bought up large parcels of land in the San Fernando Valley, located north of Los Angeles. Water from the aqueduct would make the semiarid region arable, they anticipated, but when it became clear that no such water would ever be made available outside of Los Angeles, the San Fernando landowners argued for annexation. By 1924, their successful campaign had ended up quadrupling the area of the city, the population of which in the wake of World War I was growing at the rate of one hundred thousand per year. This growth, combined with a three-year drought, severely taxed the water supply. Under worst-case conditions, ranchers in the San Fernando Valley were intercepting virtually all of the aqueduct's base flow. The city of Los Angeles sought to acquire more water rights in the Owens Valley, but angry residents, still bitter from the original conflict over the claims of the rural valley versus those of a growing city, balked, and some turned to violence reminiscent of the Old West. Among the retaliatory acts was the dynamiting of the aqueduct, which subsequently had to be protected by armed guards.

In the meantime, recognizing that the aqueduct could not supply enough water for both urban Los Angeles and rural San Fernando Valley without enormous storage capacity, additional reservoirs had been planned and designed and were under construction. In fact, between 1920 and 1926, a total of eight new reservoirs were built by the Los Angeles Bureau of Water Works and Supply, during which time Mulholland made it known that it was his goal to have enough reservoir capacity to hold in reserve an entire year's worth of water for the city. Among the additional reservoirs Mulholland planned was one that would account for about half the total water required. This dam, to be located in San Francisquito Canyon, was to be called the St. Francis.

The St. Francis dam site was chosen after inflated land values had made another location too expensive for Mulholland's tastes. In fact, he had imagined a dam in the San Franscisquito during the construction of the aqueduct. Then, a construction camp had been set up at the site to house men working on the tunnels along the canyon, the broad flat bottom of which was bounded by steep sides. Mulholland saw that a relatively small dam built where the canyon narrowed would hold back an enormous amount of water. He also recognized early on that the geology of the location called for special caution, but these conditions did not keep him from designing a dam for the site. He assumed that the buttressing effect of the dam would mitigate any slippage at the canyon walls.

Until 1923, all the dam designs that Mulholland had overseen were earthworks, large embankments whose fine-grained silt and clay cores were more or less impermeable to water. The first concrete dam built for the city of Los Angeles was the two-hundred-foot-high Weid Canyon dam, which was designed to impound the Hollywood Reservoir. It has been speculated that Mulholland decided to adopt a concrete-dam design over the clay-core type with which he was so familiar because of the limited supply of clayey materials in the sides of Weid Canyon. A year before the unique concrete dam was completed, in 1925, it was christened Mulholland Dam, a testament to the stature of the chief engineer.

St. Francis Dam was similarly designed to be made of concrete, because there was no suitable clay or silt available at the San Francisquito Canyon site. The new dam would also be a stepped concrete gravity-arch structure: Its downstream face was constructed like a wide set of steps; its material was mass (unreinforced) concrete; and it held back the water through its sheer weight pressing down on the ground, aided by an arched plan that took advantage of the water pressure behind it to compress or wedge the dam between the sides of the canyon, which served as abutments.

The original design of the St. Francis called for a dam reaching 175 feet above the San Francisquito Creek bed, which would have given it a capacity of thirty thousand acre-feet of water—that is, enough water to flood thirty thousand acres to a depth of one foot—enough to supply Los Angeles for one year. But because of increased water use by Los Angeles, before the first concrete was poured the dam's capacity was increased to thirty-two thousand acre-feet by raising its height and adding a wing dike that extended from the west abutment. Almost one year after the beginning of the placement of concrete, apparently in response to further growth in water usage, the reservoir capacity was once again increased by raising and extending the wing dike and by adding another ten feet to the dam's height, to increase its capacity to more than thirty-eight thousand acre-feet—more than 25 percent greater than the original design. The changes in its height had been made without a proportionate widening of the dam's base, but Mulholland believed the original design was still sufficiently safe. A gravity dam derives its ability to hold back water without tipping over from the width of its base, however, and so the factor of safety of the dam as built was definitely reduced by the design changes. Mulholland believed that St. Francis Dam still had a "factor of safety of three or four," meaning it was three or four times stronger than absolutely necessary.

Construction of the dam lagged that of its sister structure, Mulholland Dam, by about one year, and the successful advance of that structure must have inspired plenty of confidence in the safety and robustness of the basic design, despite some less than conservative design features. St. Francis Dam contained 130,000 cubic yards of concrete but no reinforcing steel. The main structure also lacked contraction joints, which allow concrete to crack in a controlled manner as it cures. (The grooves in a concrete sidewalk cause it to crack at the base of these reduced sections, thus keeping the predictable cracks more or less straight and hidden.) No doubt the arched nature of the dam was expected to close as much as possible any cracks that did develop. St. Francis Dam was also constructed without drainage galleries—tunnels that run through the structure to allow for inspection for cracks and sources of leakage and to provide a means for monitoring the amount of water flowing through the dam. Finally, the dam was built without cutoff walls (concrete-filled trenches designed to reduce water seepage

under the dam) or a grout curtain (a further seepage-prevention measure taken by forcing grout under pressure into holes drilled in the rock
under a cutoff wall). These measures reduce the possibility of water
infiltrating under the dam and exerting upward hydrostatic pressure,
thus making the structure somewhat buoyant. Such buoyancy provides
an uplift force that reduces the effect of the weight of the dam in keeping it in place. In extreme cases, uplift forces can cause a dam to tend to
tilt forward or slide downstream. In short, many of the design features
of St. Francis Dam, which were in accordance with standard engineering practice of the time, in retrospect contributed to making it less
watertight, less inspectable, and less stable than should have been considered wise.

By the time he was building the St. Francis, Mulholland's record of
successful dams appears to have given him a confident attitude toward
the ability of the gigantic structures to hold back the force of water, but
he had also resigned himself to the fact that some water leaked through.
Although his experience was with earthen dams, he evidently felt comfortable transferring his confidence to concrete dams, which, after all,
were made of a stronger and less permeable material.

St. Francis Dam was completed in May 1926, but months before that
date water from the Owens aqueduct was diverted into the reservoir. At
first, enough water was allowed to pass through outlets in the dam to
maintain the flow in San Francisquito Creek. Shortly after the dam was
completed, however, Los Angeles requested the appropriation also of
"flood and surplus waters" and blocked the flow into the creek. Mulholland is said to have believed that the Santa Clarita Valley ranchers
downstream could continue to draw water from their wells, apparently
not appreciating that the replenishment of the groundwater depended
upon the creek flow. An agreed-upon test release of water from the dam
showed the resulting stream to dry up within a few miles, indicating
that the water indeed was going into the ground. This incident was one
of Mulholland's few public embarrassments over water issues, but it
also demonstrated that his assertions about water flow through the
ground were not always fully informed.

A year after St. Francis Dam was completed, the level of its reservoir
reached within three feet of the crest of the spillway, which was
designed to keep water from overflowing the top of the dam. The water

St. Francis Dam, with water issuing from an outlet gate

did not reach the spillway, however, for the spring runoff ceased, and the level of the reservoir began to drop. The cracks that had developed in the dam during its filling were described by Mulholland as "transverse contraction cracks" and did not appear to alarm him. The downstream crevices were "infilled with hemp and sealed with wedges of oakum" and "backfilled with cement grout to seal off active seepage."

The next year's spring runoff caused the reservoir to fill again, this time to maximum capacity. In the meantime, new leaks developed in the dam, some manifesting themselves in springs in the foundation and others in the old cracks—through which the discharge was increased over the previous year. Still other leaks developed on either abutment of the dam and in the wing dike. Mulholland ordered a concrete pipe installed to drain water from this last leak toward the abutment of the dam.

March 12, 1928, was a windy day, and water from the reservoir was being blown in waves against the dam and over its spillways. This water naturally washed over the stepped downstream face of the dam, making it difficult to tell if new leaks were developing or old ones were growing. Full reservoirs throughout the system and seasonal runoff combined to present an abundance of water, which was allowed to flow into San Francisquito Canyon for the first time in almost two years. Earlier that day, the St. Francis damkeeper had called Mulholland to bring to his attention a new leak, one of "dirty" water, at the west abutment. Such water could indicate that foundation material was being washed out from under the dam, which could lead to it being undermined—certainly a dangerous condition. Mulholland, who claimed to have made it a practice to visit all nineteen dams under his supervision at least once every two weeks, immediately drove with his assistant to the St. Francis, where they spent two hours inspecting the dam. The "dirty" water seemed clear to them, and the dam was declared safe.

A little before midnight that same day, the dam gave way, and the contents of the reservoir inundated San Francisquito Canyon. The water rushed down the canyon, destroying everything in its path. Some large sections of the concrete dam, weighing thousands of tons, were washed as much as one mile downstream, leaving only the tall center section of the structure in place, with some other large blocks scattered nearby, mostly between the center section and what had been the east abutment. A powerhouse located about a mile and a half downstream was washed away, as were a construction camp and houses in little towns and villages in the path of the water. Hundreds of people were killed, most no doubt unsuspecting as they slept. The official death toll was in excess of 430, but the actual total is debated to this day.

According to *Engineering News-Record,* which had long ago established its reputation for accurate and incisive reporting on the failure of structures, it was "the first time in history a high dam of massive masonry" had failed. The disaster was compared to the Johnstown Flood of 1889, which had claimed more than two thousand lives and was once considered "the worst in history resulting from failure of man-made structures." However, the magazine declared, "the washing out of an old neglected earth dam was not an engineering tragedy" so much as a case of carefree modifications and poor maintenance by the hunting and fishing club that had patched an abandoned Pennsylvania state-canal-system reservoir to make a recreational lake. The failure of the St.

Aftermath of the St. Francis Dam failure

Francis Dam was indeed different, for here it was assumed that the latest engineering materials, design philosophies, construction techniques, and operational procedures were overseen by an engineer with an impeccable record of success. Mulholland's work was not without critics, but in the business of holding back vast quantities of water, he had been able to answer their fears. At least in the minds of those who were in the position to give the go-ahead for such a great project, a concrete dam built by Mulholland would certainly be stronger than the water that pushed against it. But, as an editorial in *Engineering News-Record* put it, "Men have always been in awe of these vast forces, and often has bitter protest been made against the erection of a dam above populous communities. In every instance engineering science answered the protest and gave assurance that the waters would be safely controlled. The destruction of the St. Francis dam challenges that assurance."

A great failure is the perfect counterexample to a hubristic hypothesis. William Mulholland and his staff had evidently so gained confidence in their mastery over the great hydraulic forces pent up behind the successful dams they had built that they began to construct them with less and less attention to detail, especially the all-important local detail of the geology underlying the site. Or, if they did pay close attention to it, they missed some key elements of its character in San Francisquito Canyon. Mulholland admitted as much when he testified at a coroner's inquest:

> I have had rather more experience than most engineers in building dams. I have built nineteen of them and took the normal precautions, judging the formation the best I could, the hazards which the dam must be exposed to and all the things that relate to the continuous safety of the dam. We overlooked something here. This inquiry is a very painful thing for me to have to attend, but it is the occasion of it that is painful. The only ones I envy about this thing are the ones who are dead.

More than a dozen official boards and commissions were appointed by various California offices and interested parties, ranging from the state governor to the Los Angeles County district attorney, to investigate the St. Francis Dam disaster. Within a month or so of the incident, studies were made, witnesses were interviewed, hearings were held, and a half-dozen reports were filed. All of these identified the foundations of the dam as having been inadequate, but there was no unanimity over the exact triggering mechanism for the catastrophe. To this day, the hypotheses put forth in the reports remain unconfirmed in any final sense, however, for the structure no longer exists.

The official failure reports dismissed early speculations about an earthquake or explosion causing any initial breach. They also confirmed that the concrete was of sufficient strength and that the design of the superstructure was in accordance with commonly accepted engineering practice. (The failure appears not to have had any significant effect on the structural design of concrete dams at the Bureau of Reclamation, which had succeeded the Reclamation Service in 1923.) However, the reports condemned the foundations underlying the St. Francis Dam, a subject that in general was incompletely understood at the time.

Among the things that were pointed out in the various reports were that the conglomerate material under the upper-right abutment structure was found to be held together with clay that softened when wet and furthermore contained numerous fissures filled with gypsum, which is dissolvable in water. Any water flowing through this material would in time carry out gypsum, which would not be easily seen, for the water would appear to be clear. When sufficient gypsum had been dissolved and the foundations were sufficiently softened, the dam would settle unevenly, eventually crack, and finally have no ability to hold back the water behind it, which would rush freely downstream.

The reports were in general agreement that it was the west side of the dam, where the underlying rock was as described, that gave way first. However, within weeks of the disaster there were also indications that the east side of the dam may have failed first. Geologist Bailey Willis and the father-and-son engineer team of C. E. and E. L. Grunsky pointed out that the east side of the dam had abutted the steep slope of the canyon, which consisted of parallel layers of schist that were faulted. It was clearly the site of an ancient landslide, something that Mulholland evidently did not appreciate. When water backed up behind the dam, it began seeping into the faulted schist of the abutment, eventually leading to ground movement, which in turn led to the movement and fracture of the dam itself.

The landslide that was evident at the site of the failed dam was generally assumed to have occurred after the dam was breached. More recently, J. David Rogers, who had the benefits of computer analysis and simulation, has argued that it was in fact a landslide of the east canyon wall that triggered the dam failure, with the rush of water scouring the dam's foundation and causing the dam to lean. This in turn opened up preexisting cracks all along the arched structure. According to Rogers's recent studies, the landslide dumped as much as one million cubic yards of weathered mica schist—as much as five times the weight of the dam itself—which "created an outpouring flood wave supercharged with sediment." If the density of such sediment was sufficiently greater than that of water, Rogers believes that it could effectively have made large sections of the dam buoyant enough to be pushed and tumbled down the canyon, where they were found in the wake of the ensuing flood. This failure hypothesis, like all the others, remains ultimately unprovable in any incontrovertible sense, of course, but it highlights the complexity of anticipating the forces on a concrete dam structure.

One vocal opponent of the view of Rogers and others about the causes of and blame for the failure of the St. Francis Dam is Donald C. Jackson, a historian who has studied contemporary dam construction. Jackson believes that the official panels that reported so quickly after the disaster were not thorough or candid enough in their assessments of the causes of the failure. Rather than attributing the disaster to weak foundations leading to the undermining of the dam, Jackson lays blame directly on the design of the dam structure itself—and hence on Mulholland. According to Jackson, the danger of uplift due to pressure from water seeping under an insufficiently designed dam was recognized by engineers in the late nineteenth century. This upward pressure, combined with the accompanying effect of a lessened resistance of a dam to being pushed downstream by the water behind it, was believed to have been the cause of the 1911 failure of a dam in Pennsylvania. To avoid such incidents, engineers such as John R. Freeman, the East Coast engineer who had consulted for Mulholland on the Los Angeles aqueduct, promoted grout curtains and drainage systems for dams potentially subject to seepage. Jackson has documented the extensiveness of discussions of uplift in the contemporary engineering literature and has provided evidence of dams designed to obviate the phenomenon. The St. Francis Dam did have some uplift-relief wells located under its main channel section, but there were no similar precautions taken against uplift for the abutments of the dam, which were located on steep slopes.

Regardless of the exact mechanism by which the St. Francis Dam cracked and gave way, William Mulholland took full responsibility for the disaster. On the witness stand, he admitted that he could not explain the failure and that something must have been overlooked. Overlooking something is, of course, always a danger in the design of large engineering systems, and it is precisely why the opinion of independent experts is sought during the design stage. In addition to the technical error of siting the St. Francis Dam on poor foundations, its collapse was blamed on the "human factor," which manifested itself in the fact that "engineering work in the Bureau of Water Works and Supply always had been dominated by one man, the chief engineer, who took upon himself, in this case at least, entire responsibility, sought no independent opinions and adopted technical policies based on his unconfirmed judgment alone." Since "higher officials had absolute confidence in Mr. Mulholland," outside opinion was not sought, and "there was no intervention from above." The plans for the dam were not challenged because "Mr. Mulholland was personally overseeing the work."

No engineer should have such hubris as to think that his past successes are sufficient to guarantee the success of his next project. Each new undertaking rests on a new foundation, the hidden faults of which may or may not be within prior experience. When all dams and the foundations upon which they rest begin to look alike to an engineer like William Mulholland, he himself should question his own expertise. As *Engineering News-Record* put it just two months after the disaster, "Had the plan of construction used for the St. Francis Dam been brought forward by some comparatively inexperienced engineer, or had the work been done by contract or under any other condition that would naturally have brought independent engineering opinion into the case, it is highly probable that some modified plan would have been substituted and the disaster avoided." Experience alone is not always the best teacher.

Three Gorges Dam

The Yangtze is the third-longest river in the world, behind the Nile and the Amazon. Originating from the 5,800-meter-high Mount Tanggula in the Tibet Plateau known as the roof of the world, the Yangtze follows a roughly west-to-east route for more than 5,500 kilometers, turning quite sinuous at times, before emptying into the East China Sea at Shanghai. The river has 3,600 tributaries and drains almost two million square kilometers, which amounts to almost 19 percent of China's land area. At Yichang, some thousand kilometers from its estuary, the Yangtze has an average discharge of almost fifteen thousand cubic meters per second.

During flood season, the water level in the river can rise as much as fifteen meters, affecting fifteen million people and threatening 1.5 million hectares of cultivated land. Historic floods have been devastating. The flood of 1870 is still talked about along the middle reaches of the river, and the one in 1954 inundated three million hectares of arable land and claimed thirty thousand lives. Altogether in the twentieth century, as many as a half-million people may have died in the Yangtze's floodwaters.

The Yangtze also has some of the most beautiful scenery in the world, especially in the region known as the Three Gorges, with spectacular cliffs and steeply sloping mountains rising as high as 1,500 meters. Interspersed with gently rolling hills and long, sloping riverbanks, the gorges have been compared in majesty to the Grand Canyon.

Cruising the river through the Three Gorges is considered a classic travel experience, as each bend in the river reveals a new perspective on the marvels that geological change has wrought.

Balancing the desire to preserve the river in all its natural glory against that to tame it to control flooding, generate power, and provide more reliable shipping conditions presents a classic dilemma involving engineering and society. When nationalist leader Sun Yat-sen proposed a Three Gorges dam in 1919, the ecological costs were overshadowed by the economic benefits for China. In the mid-1940s, a preliminary survey, along with planning and design efforts, was carried out by the U.S. Bureau of Reclamation under the direction of John Lucian Savage, designer of the Hoover and Grand Coulee dams. In his exploratory role, Savage became the first non-Chinese engineer to visit the Three Gorges with the thought of locating an appropriate dam site. Savage's work is the likely inspiration for John Hersey's novel *A Single Pebble,* the opening sentence of which is, "I became an engineer." In the story, the unnamed engineer travels up the Yangtze in a junk pulled by trackers in the ancient and, once, the only way to make the river journey.

Chairman Mao Zedong was a staunch supporter of a Three Gorges dam, which he felt would provide a forceful symbol of China's self-sufficiency and ability to develop its resources without western aid. As early as 1953, Mao expressed his preference for a single large dam rather than a series of smaller ones, and he suggested that he would resign the chairmanship of the Communist Party in China to assist in the design of the project. Mao's poem about being at ease swimming across the Yangtze reflects on how all things change, like the swift river and the gorges through which it flows. In the poem, he knows that the Goddess, a prominent peak in the middle reaches of the Three Gorges, will marvel at the accomplishment of a dam.

In 1992, the Chinese government announced officially its determination to tame the Yangtze with what would be the world's largest hydroelectric dam, ultimately to be fitted with twenty-six generators rated at seven hundred megawatts each. The total of 18,200 megawatts is equivalent to the output of approximately fifteen of the largest nuclear power plants operating in the world today or of a like number of coal-burning plants. Since one of China's most pressing environmental problems is pollution from burning fossil fuels, the prospect of a hydroelectric dam cleanly generating about 10 percent of the country's power was very appealing to the Chinese leadership. In addition to providing flood control and power generation, the dam will open up the Yangtze as far upriver as Chongqing to ten-thousand-metric-ton ships, providing an opportunity for China to develop container ports almost two thousand kilometers inland. This purpose goes hand in hand with China's plan to become a full partner in world trade operations.

For all of its practical benefits to China, the Three Gorges dam project was opposed by numerous groups both domestic and international. Especially vocal were human-rights advocates, environmentalists, and historians. Among the most persistent opposing voices was that of Dai Qing, who was educated as an engineer but became disillusioned during the Cultural Revolution and finally turned to investigative journalism. Her 1989 book, *Yangtze! Yangtze!*, was highly critical of the idea of a Three Gorges dam and led to her temporary imprisonment. That book and her subsequent one, *The River Dragon Has Come!*, published in 1998, state the fundamental case against the project but seem to have had little, if any, effect on the progress of the dam.

When fully complete, the Three Gorges Dam, which will stretch about two kilometers across the Yangtze at Sandouping, will be 185 meters high and will create a reservoir six hundred kilometers long, reaching all the way to Chongqing. The filling of the reservoir was expected to displace on the order of one million people, inundate almost fifty thousand hectares of prime farmland, submerge archaeological treasures, and forever alter the appearance of the Three Gorges.

The project has been described as "perhaps the largest, most expensive, and perhaps most hazardous hydroelectric project ever attempted." Vocal protesters and international politics no doubt influenced the World Bank's refusal to finance the project. Bowing to pressure from environmental groups, the Clinton administration opposed competitive financing through the Export-Import Bank, effectively discouraging American companies from participating. The Chinese government has nonetheless been resolute.

As in all large dam projects, picking the site was of fundamental importance. Of fifteen locations seriously considered, the final choice was made on the basis of geological foundation conditions and accessibility to construction equipment and materials. The chosen dam site is twenty-eight kilometers upriver from Yichang, near a location where the Yangtze runs wide between gently sloping banks that provide staging areas for the construction project. The geology in the area is ideal, in that it is underlaid with solid granite for some ten kilometers surrounding the dam site, providing a stable construction base. A small island in the river proved to be an advantage in diverting the flow to one side while construction began on the other.

The project was planned to be completed in three stages. Phase 1, stretching from 1993 to 1997, consisted of building cofferdams within which the river bottom could be excavated and the foundation of the dam begun. A temporary ship lock was also constructed during this phase, in order to allow shipping to pass the construction site throughout the project. Phase 2, extending from 1998 to 2003, involved building a good part of the dam proper. Concrete was poured twenty-four hours a day to complete the spillway of the dam, the intake portions designed for power generation, and the initial stage of the power plant itself. At the end of Phase 2, the reservoir began to be filled so the dam could begin to generate power, thus producing revenue to fund the final phase of the project. Phase 3, planned to stretch from 2004 to 2009, involves the completion of the dam across the river, including additional powerhouse units. According to the China Yangtze Three Gorges Project Development Corporation (CTGPC), the government-authorized entity created to own the project, construction remains on schedule.

In the fall of 2000, under the sponsorship of the People to People Ambassadors Program, I led a civil-engineering delegation from the United States to visit the Three Gorges dam construction site and talk with Chinese engineers about the project. We had the opportunity to see firsthand the scale and technical nature of this gargantuan engineering project and to visit the areas along the Yangtze that will be permanently altered by the creation of the reservoir. The delegation saw towns that would be submerged and from which so many people would be displaced. We also had an opportunity to experience China in this time of rapid emergence as a full player on the world economic scene.

Our delegation consisted of forty engineers. They were mostly civil engineers, some with extensive experience in dam construction and power generation, but there were also a number of electrical and

Rendering of the Three Gorges dam project

mechanical engineers, among others, as well as a geologist, reflecting the inherent interdisciplinary nature of large engineering projects. About twenty-five guests traveled with the delegation. Most were spouses of the delegates, but there were also a half-dozen sociologists, who were most interested in the relocation problems associated with the project.

The conventional wisdom in the United States about the endeavor was that it was technologically risky, environmentally unsound, sociologically devastating, and economically unwise for China to have undertaken. Thus, the overall view of the Three Gorges Dam held by most Americans was that it was ill-advised at best and a disaster in the making at worst. Different members of the delegation took different preconceptions with them to China, and some brought home altered perceptions.

The delegation assembled in Los Angeles in mid-November for a predeparture briefing, much of which concentrated on practicalities of traveling in China. We were warned against drinking the water, told to have local currency for airport taxes, cautioned against setting too high expectations of amenities, and instructed to observe professional protocol.

A five-hour layover in Hong Kong gave some of us an opportunity to ride the new Airport Express into the city that is now part of but still apart from China. The train is state of the art, with many seats facing video monitors on which riders can call up train routes, weather, stock quotations, news, and other features. We engineers, however, focused mainly on the view outside the train, struck by the volume of construction and the size of the ship-container handling facilities. After a twenty-five-minute ride, we emerged from the subway in the middle of Hong Kong Island, where we explored some most impressive office buildings before returning to the train station for the trip back to the airport. The train was crowded with commuters, many of whom boarded at Kowloon, one of only two stations between Hong Kong's financial district and the airport. The station is air-conditioned against the Hong Kong heat and humidity, and to make that practical the station platform is separated from the tracks by a glass partition. The train stops precisely at the doors in the partition, and riders with luggage, whether going to or coming from the airport, can find an abundance of baggage carts conveniently dispersed along the platform.

The world of difference between Hong Kong and the interior of China was emphasized by the fact that our flight from Hong Kong to Wuhan was classified as international. Wuhan was created in 1950 out of the merger of three cities separated only by the Yangtze and Han rivers, a region rich in Chinese history. The consolidated city is located midway on a north-south line between Beijing and Canton and an east-west line between Shanghai and Chongqing. In contrast to the bright stainless-steel ambience of the new airport at Hong Kong, the dated one at Wuhan was tiny, dingy, and drab. Wuhan, one of China's industrial cities, has a population of about four million. It was our point of entry into China's interior because it is on the Yangtze River and conveniently connected by a modern toll highway to Yichang, headquarters of the CTGPC and just downstream from the construction site.

Although a straight shot on a new superhighway, the bus ride to Yichang was to take about four and a half hours, so we spent the night in Wuhan. This important river port evokes nineteenth-century technology as much as Hong Kong does twenty-first. Wuhan has its buses and cars, the increase of which throughout China is creating enormous traffic and pollution problems, and its better hotels, complete with CNN. But the enormous reliance of the people on muscle power harks back to the earlier century. Myriad bicycles have their dedicated lanes, and they carry goods that in America would be found in pickup trucks and delivery vans. It is common to see bicyclists struggling up the slightest incline under a load of reinforcing steel or plastic pipe that extends several meters ahead and behind the bike. Smaller loads, though not always smaller by much, are carried in bundles hung from the ends of bamboo poles balanced on the shoulders of bearers, who trek along among the bicycles.

Our one excursion in Wuhan was to a restored ancient hilltop pagoda, perhaps the city's most famous tourist attraction. On the ride to and from the Yellow Crane Tower, which was apparently named after a mythic bird—there are no cranes colored yellow in China, our tour guide informed us—we crossed and recrossed a road-and-railroad bridge spanning the Yangtze. Through the haze with which we would become quite familiar, we could glimpse a newer cable-stayed crossing in the distance, one of the many newly constructed modern bridges we would encounter as we crisscrossed the country.

The next day, the bus ride to Yichang was through primitive farm-land. In sharp contrast to the new cars and buses traveling the highway on which we rode, the farms showed no sign of mechanization. Those farmers who did not walk behind a water buffalo worked bent over in their fields. During harvest, farmers lived in the tents and tiny shacks that abound beside the fields. Clusters of small, run-down farmhouses marked simple villages, with virtually all buildings oriented with their entrances facing south, in the tradition of much-grander Chinese houses. Although in the days of collective farming there was some machinery, our guide told us, that has not survived into the present era, when smaller plots of land are worked by individual farmers. But the state still owns the land in China.

Since it was late in the year, there were few crops in evidence. The clearly irregular fields followed the contours of irrigation ditches, and some worked-out fields were excavated deeper than their neighbors to allow for fish farming and lotus cultivation. Hubei Province's great Jingbei Plain west of Wuhan is extraordinarily flat, and after several hours' riding we had become so accustomed to the flatness of the land that the sudden appearance of hills with terraced fields jolted many of us out of a torpor.

The presence of hills soon yielded to mountains, which signaled our approach to Xiling Gorge, the most downriver of the Three Gorges and thus often referred to as the third gorge. The twists and turns of the highway through the mountains caused the city of Yichang to appear as suddenly as new stretches of river would when we would sail through the gorges a few days hence. The most prominent building to first come into view in Yichang was the modern China Telecom Building, the city's tallest. It and the headquarters building of the CTGPC dominate the skyline of hilly Yichang, which has come to be known as "dam city" and "electricity city," in recognition of the many hydroelectric power plants in the area.

Before leaving the Yichang area, we visited the Gezhouba Dam, completed in 1981 and a prototype of sorts for the Three Gorges Dam. Located about thirty kilometers downstream from the construction site, this hydroelectric dam has all the features of the larger structure. In particular, it has sediment-control gates, which when opened scour out accumulated sand and silt from behind the dam and distribute it downstream. The issue of accumulating material behind the Three Gorges Dam is one of the objections raised by opponents, who argue that in time the reservoir will fill with silt and become unnavigable. Impounding the silt behind the dam will also deprive the agricultural land downstream of natural replenishment. The reportedly successful operation of Gezhouba Dam, however, appears to have allayed immediate concern about silt, at least among engineers.

It was dark when we left Gezhouba Dam and boarded the buses for the ride to Sandouping, the base town for the Three Gorges dam project. We could not see the terrain through which we were riding, but the grades of the hills and the rock slopes visible in the bus's headlights made it clear that we were traveling through rough territory. The road was new, constructed to serve the project site, and it led to a heavily guarded check station. At one point, we passed through a tunnel about two kilometers long, suggesting that the mountains above us were too high or steep to put a road over. After perhaps forty-five minutes of riding without seeing any significant number of lights, we came upon San-douping, a small town by Chinese standards but a bustling center beside the Yangtze. Our hotel was a relatively new high-rise. From its windows we could see the outline of lights on the cables of what must have been a major suspension bridge, suggesting that we were beside the river.

In the daylight, we would learn that the attractive and graceful structure was the Xiling Bridge, a major span with well-proportioned white concrete towers and a strikingly slender red roadway. The bridge seems to herald the dam, which was only a short drive from our hotel. On the way to the construction site, we passed numerous warehouses and dormitories for workers. These latter, we were told, will be converted into tourist accommodations. Dominating the route to the construction site was a large pit where granite was being crushed into pieces of aggregate for the concrete. A system of conveyor belts carried the stone over and along the road to the concrete plant. This location thus provided not only a solid foundation for the dam itself but also a convenient source of one of the principal materials for it.

The construction site was so large, extending well over a kilometer out from the riverbank and into the riverbed, that it was hard to encompass in a single view. Perhaps the dominant first impression was the countless number of tall construction cranes, literally countless because they blended into one another and disappeared behind one another. Our guide told us that we were finally seeing a real yellow tower crane, as opposed to the mythical Yellow Crane that we had puzzled over in Wuhan.

The first stop on the site was at the location of the locks. Twin pairs of five locks were being carved out of solid granite and lined with concrete. They will carry ships and barges in stages through the difference in water level behind and in front of the dam. Viewed from near the bottom, the scale of this one aspect of the Three Gorges project was enormous. I certainly had never seen anything like it, and I imagined that it rivaled even the construction of the individually larger locks of the Panama Canal. Workers at the bottom of the man-made granite box canyons looked minuscule, and it seemed impossible that these locks were blasted out of the granite in only a few years' time. Our guide told us that the Chinese calligraphy atop a nearby promontory motivated the workers to keep at the task with "first-class management, high-quality worksmanship, first-rate construction."

For American engineers, one notable feature of this Chinese construction site was the freedom with which we visitors were allowed to move among the piles of construction materials and debris. Such traipsing around (and without hard hats) is unheard-of at construction sites in the United States. Only a simple railing with wide openings separated us from a thirty-meter fall into one of the ship-lock excavations, but no one seemed to be bothered by the proximity to the precipice.

After spending some time at the locks, we were driven over to the dam proper, which we viewed head-on from its downriver side. The scale of this part of the project was even grander than that of the locks, for it rose higher into the air and stretched over a kilometer wide before us. Under construction to our left was the spillway, with one section of it raised to the dam's final height, giving a sense of how the finished structure would loom over this part of the river. To our right was the power-plants section, which would hold fourteen hydraulic-turbine generator units capable of generating 9,800 megawatts of power when this portion of the dam was completed. (The remaining 8,400 megawatt capacity will not be realized until the third stage of the project is completed and all potential generating capacity is in place.)

Behind us stood the batch plant, where concrete was being mixed constantly for the twenty-four-hour-a-day work schedule. One of the major considerations in placing concrete in such a massive structure is how to dissipate the heat of hydration that is generated in the concrete as it cures. The concrete, which experiences thermal contraction as it cools, can develop cracks. Taking the heat away in a controlled and timely manner obviates this unwanted behavior. At the Three Gorges Dam, the thermal problem was being handled in several ways. As with Hoover Dam, cooling pipes were imbedded into the concrete to carry away some of the heat. A certain amount of undesirable heat was itself eliminated at the source by mixing and placing the concrete at the lowest temperature possible. This was accomplished through cooling the aggregate by blowing cold air over it, by using ice water in the mixing process, and by insuring that no concrete comes out of the batch plant at over seven degrees Celsius. The measures appeared to be working.

Only one significant crack had appeared in the part of the dam in place, and the Chinese engineers seemed confident that it was satisfactorily repaired.

After the dam itself and the tower cranes—red, white, *and* yellow— the next most prominent feature of the construction site was the conveyor-belt system that rose up to great heights on temporary concrete columns. The conveyor system to deliver the concrete was a critical component of the job, for the rate at which concrete is placed largely determines if a project can be kept on schedule. Unfortunately, shortly before our visit to the site, an accident involving one of the conveyors had killed some workers. Before that accident, we were told, the safety record of the project had been excellent. At the time of our visit, the conveyor system was not operating at the desired capacity, which irritated the Chinese, and local papers were carrying stories of a lawsuit against the conveyor company for breach of contract.

While at the construction site, it was hard not to be awed by the enormity of the project and the confidence of the engineers working to hold back the legendary Yangtze, building on their experience with Gezhouba Dam and the many other flood-control and hydroelectric projects completed throughout their country in recent decades. (A late-twentieth-century survey by the World Commission on Dams found that 46 percent of the world's forty-five thousand large dams were located in China. It also reported that, although they have contributed significantly to human development, dams have also been the cause of considerable social and environmental damage.)

The convincing official arguments that the Chinese put forth about the multifarious good that the Three Gorges Dam will bring to their emerging economy impress visitors from a country that is what it is today in part because its engineers also tamed great and scenic rivers like the Colorado and the Snake. The preconceived opposition to the Chinese project as being merely irresponsible and anti-environmental, a view that some members of our delegation brought with them from America, was allayed as we stood before this monument in progress.

The Three Gorges Dam is a construction project comparable in human resolve to those of the Egyptian pyramids, the Great Wall of China, and the Panama Canal. For all of the human tragedy associated with such megaprojects, tourists flock to them to see the accomplishment. Spectators are attracted to the scales and the stories of these achievements and to the recognition that they say something about the human aspirations of our ancestors even as they were being inhuman. Today, in the geological interlude between the east and west parts of the Xiling Gorge, the Three Gorges Dam presents a tangible expression of the assertion of the People's Republic of China that it is the equal of any country on earth. During debates over China's role in the world economy and its political place among the superpowers, it has moved forward with dam building, bridge building, road building, and city building on a scale that is reminiscent of early-twentieth-century America.

In a modern exhibition building barely one kilometer from the Three Gorges construction site, the case for the dam is made in what might be seen as a western-style public-relations effort, were it not for the exclusive use of Chinese in the display captions. Although the exhibit is intended as an introduction to the construction project, we visited it after we had viewed the site because a high-level official government delegation had preempted us. Once that delegation had gone, we assembled around a five-by-ten-meter model of the dam and its environs as if around a conference table. A senior engineer, through an interpreter, gave a succinct and effective introduction to civil-engineering delegates and guests alike. Both the engineer and the translator used miniature laser pointers, no doubt made in China like virtually every other artifact that we encountered there. The country seems soberingly self-sufficient in everything from financing a mega-project to making trinkets.

The exhibits in the museum-like hall addressed all aspects of the Three Gorges project, including the technical details of transmitting the power over high-voltage direct- and alternating-current transmission lines to places as far away as Shanghai. The engineer discussed environmental impact, concluding that it was of acceptable proportions. Many significant archaeological treasures that would otherwise be inundated were being relocated to museums, and people who were to be flooded out of their ancestral homes were to be resettled. From the American perspective, resettlement is one of the most controversial aspects of the dam, but in China it appears to be treated as just another cost-benefit decision that had to be made and but another sociotechnical problem to be overcome.

Our delegation had the opportunity to see firsthand one of the new towns already built to accommodate people who would be displaced by the reservoir. The old town of Zigui is located about fifty kilometers up the Yangtze from the dam site and, with the dam completed and the reservoir filled, would be underwater. This necessitated relocating it. The new Zigui town is situated near the dam on a hill that overlooks the reservoir, which provides scenic and recreational resources for the townspeople and any tourists that it attracts. (After the completion of the dam, its environs were expected to be as inundated with tourists as the Three Gorges would be with water, and there was much anticipation that this would be a shot in the arm for the local economy. So great was the expectation of major tourism that the banks of the river beneath the dam and roads surrounding the new Xiling Bridge were being landscaped in a style reminiscent of Berlin or Paris.)

Zigui is among the first of the new towns to be occupied, with an initial population of approximately thirty thousand and the prospect of perhaps doubling before too long. At the municipal building, we met the chief planner, who assembled us around a model that showed the planned town to be laid out beside the completed reservoir, with a gently sloping hill dividing the town into distinct sections. While residential parts appeared to include plenty of green space, most of the apartments are located in clusters of seven-story buildings, reminiscent of "the projects" in large American cities. Some of the apartments no doubt command spectacular views, but since the buildings do not have elevators the top floors are less desirable than the middle floors. When asked, our guide explained that apartment buildings were assigned to groups of workers associated with a factory team. Their manager was responsible for allocating the individual apartments within the building.

Even with this lack of control over their choice of housing, young Chinese apparently believe that relocation was an opportunity for them to advance their standard of living. For those young people from old towns with few modern amenities and with cramped quarters, the prospect of more bedrooms, bathrooms, and modern appliances made the disruption of their lives worth it. For the older generations, who have lived their entire lives in familiar surroundings, the prospect of moving as many as fifty kilometers away, even if into a large, modern apartment, was not so appealing. It seemed clear to many of us that the tensions that were likely to be created between the generations in this society, where the old have for so long been revered, may create unanticipated problems for relocation efforts and for China.

Shortly after returning to Sandouping, we boarded the *Snow Mountain* for our trip up the Yangtze. The riverboat was not large by cruise-ship standards, but it appeared to be trying to emulate them in its amenities: beauty salon, massage parlor, bar, and show stage. As soon as the *Snow Mountain* left the dock and sailed to the middle of the river, we gathered on the forward deck to admire the Xiling Bridge, which loomed ahead of us like a ceremonial gate. We now could see clearly that the nine-hundred-meter-span suspension bridge has side spans that are supported from beneath on the riverbank. This obviates the need for the suspension cables to support the side spans, and so the taut cables above them have a rigidly straight profile that directs the eye to the granite into which they are anchored directly. The absence of massive concrete anchorages emphasizes the sleekness of the bridge. We looked up at the underside of the deck as we passed beneath the bridge, confirming that the structure is a box girder.

The Xiling Bridge had so captured our attention that it was only after we had passed under it that we realized what a spectacular view of the Three Gorges construction site lay before us. From our midriver vantage point, we got a much better appreciation of the dam's enormous reach. Like those along the skyline of a great city experiencing a building boom—a common case in cities throughout China—the tower cranes dominated the top of the view. From our downriver approach, we could see the circular falsework in place for forming the great penstocks that would carry water from the reservoir to the turbines at the bottom of the dam. As the *Snow Mountain* cruised upriver toward the diversion channel to the left of the rising concrete megalith, we got an appreciation for the force of the river. Confined to a channel barely one third the width of its natural course, the river rushed past us with a great ferocity. (What would it be like during flood season?) As the riverboat labored against the current, we looked up at the cross section of the dam as if at a skyscraper. A worker framed by one of the inspection tunnels gave a sense of scale to the structure. No matter from what perspective we looked at the dam in progress, it promised to be a masterpiece of engineering, a wonder of the modern world. Huge red Chinese characters on placards high on a hillside proclaimed, "Build the Three Gorges, Develop the Yangtze."

As we passed through the diversion channel and proceeded around the construction site, the new town of Zigui was visible on our left high on the hill above, giving us a measure of the limits to which the water would rise in the reservoir. Behind us we could see the dam from upstream, and as it receded into the mist its immensity was dwarfed by perspective and the mountain peaks that loomed ahead.

The fabled river is full of turns as it wends its way through the Three Gorges, with seldom a long open stretch ahead by which to judge what is coming. This made for constantly changing views of the river and the gorges, and thereby of our imaginings of the future reservoir. A short way upriver, around another bend, we entered the west part of Xiling Gorge. Few members of our delegation could have been prepared for what we saw. Travel guides, postcards, and souvenir books had given us previews of the scenery, but none had prepared us for the sense of insignificance we experienced as our boat wended its way among the granite cliffs and peaks. It was some time before any of us even picked out a single road, high up on the mountainside, looking more like a footpath than a highway.

Stark signs planted high on the slopes marked the 135-meter elevation that the water would reach on the first filling of the dam; others marked the 175-meter elevation, the height to which floodwaters will eventually rise. These signs, the occasional bridge or tunnel (which high mountain roads demanded), and the abandoned old town of Zigui were the few indications that a human hand had touched this part of the world. Old Zigui was nothing but shells of colorless buildings, their windowpanes and sashes removed before the coming flood. Towns like Zigui would be leveled and many of their parts carted off before that happened, we were told, both to recycle materials and to reduce the sources of pollution to the reservoir's water.

As we continued our journey up the Yangtze, we observed other towns in various stages of abandonment. Sometimes a new white town was under construction just uphill from the elevation markers. Some-

Bridged tributary to the Yangtze, before flooding

times a new town's tiled buildings glistened in the sunlight across the river from the dark-matte surfaces of the old. The relocated people will be able to watch with mixed emotions from their new apartments— many no doubt fitted with satellite television receivers and modern plumbing—as their old town is dismantled and their old land sub- merged under the rising waters. How such sights will affect them is hard to imagine.

Just before midnight, the *Snow Mountain* dropped anchor, so that we would be able to appreciate our entry into Wu Gorge, the middle of the Three Gorges, at dawn. A short way into Wu Gorge, we tied up at the town of Wushan, built on the side of a mountain. ("Wushan" was translated as "Witch Mountain" by our guide.) The road was reached via a long flight of stairs, a typical arrangement along the Yangtze, where the water level can vary by thirty meters or more from low water to spring-runoff height. The buses carrying the delegation strained at the steep climb along the town's main street, which was very narrow and lined with small shops of all descriptions. At a wye in the road, a group of men stood with bamboo poles as if awaiting orders to fight. The guide told us this was a *bang bang jun,* variously translated as a "stick army" or a "help, help army." When in use, the bamboo poles are balanced on the shoulder, and cords at each end support baskets, bags, or any commodity in need of transportation. Throughout China, we saw this method used for carrying everything from groceries from the market to construction materials at a job site. Here in Wushan, there were also large trucks being used to haul larger quantities of construction materials higher up the mountain, to the site of the new city.

After passing over the hill on which Wushan is built, the bus carried the delegation down toward the Daning River, a tributary of the Yangtze. We were headed for a landing paved with excursion boats ready to take tourists up through what are known as the Lesser Three Gorges. The entrance to these gorges is framed dramatically by Longmen Bridge, a concrete arch structure high above the river. This classically proportioned crossing will survive the flooding, but only by a small amount, thus changing the appearance of this dramatic entrance to some of the most beautiful scenery we experienced on the entire trip. The Lesser Three Gorges stretch some fifty kilometers up the Daning and are separated by interludes of lush farmland on gentle hills. The river itself is full of shoals and rapids, and getting the excursion boats upstream was not easy.

Each boat had two pole men riding on the prow. We imagined they must be the descendants of trackers, the men who hauled boats up the Yangtze and its tributaries by pulling like draft animals on ropes. Trackers sometimes had to wade through the rapids, sometimes claw along the rocks, and sometimes crouch along pathways carved into the mountainside. Along the narrower Daning, it was possible to see up close the square recesses carved into the rock in "ancient times," a term that in modern China appears to refer to periods as late as the end of the Ching dynasty—that is, 1911. The recesses are believed to have held square timbers cantilevered out over the river, across which planks were laid to form elevated paths for local residents to move about and trackers to pull boats during the flood season. These and other artifacts were threatened by the future reservoir.

Our excursion boats did not need trackers in the traditional sense, of course, but the pole men on the prow were necessary to get the motorized craft through some of the trickier shoals. At times, the boat came to an almost complete stop as its engine strained to hold steady in the current. As the flat bottom scraped across the well-rounded rocks and pebbles, the men on the prow used their steel-tipped bamboo poles to guide the boat through the shallowest of channels and to provide some pushing power to assist when needed in gaining forward progress. At one point, where the boats were clearly going to be moving at a snail's pace, local boys waded into the water offering brass bells and other trinkets to the tourists. As we had learned early on, no hawker's price is fixed in China, and bargaining is sport for both the seller and the buyer. After one of our delegation had bought a bell for a dollar, the boys surrounded the boat clanging their bells for more dollars. Many of the tourists succumbed to the temptation, making it more difficult for the pole men to drive the boat forward. Working as hard as any tracker must have, the pole men became frustrated and began swatting their poles at the boys to find room to drive the spear ends into the gravel without stabbing a bare foot. It was only after considerable effort and what must have been fierce cursing in Chinese that the boatmen extricated the craft from the shoal and the hawkers, and the excursion continued upstream.

The trip upriver took a couple of hours, at least, but few of us were keeping track of time. The return trip with the current was naturally much quicker. That is not to say that the way was easier, for the shoals were still shallow, and one of the pole men had to use a long rudder fitted to the bow to guide the boat into what channels there were. At one point, the boat scraped to a complete stop, having missed the proper channel, and had to be ungrounded with considerable difficulty and redirected.

Back on the *Snow Mountain*, the ride was naturally more relaxed. The riverboat entered the third gorge, Qutang Gorge, shortly after lunch and continued toward our ultimate destination on the river, Chongqing. When the Three Gorges Dam is completed, boats much larger than our riverboat will routinely make the journey up the Yangtze from Wuhan, long the terminus for oceangoing vessels, to Chongqing, a distance of almost one thousand kilometers, with relative ease. Chongqing is known locally as "furnace city," because of the summer heat and humidity, and "fog city," for reasons that were immediately obvious. The Chinese claim that it is the largest city in the world, although the population of thirty-one million includes the entire municipality, which in conjunction with the Three Gorges dam project stretches quite a ways down the Yangtze and is linked governmentally directly to Beijing. Known to many westerners as Chungking, Chongqing is rich in history. Located where the Jialing River flows into the Yangtze, the city was the capital of China from 1937 through 1943 and was headquarters of the Chinese nationalist armies. It was bombed by the Japanese during World War II and served as an American air base in 1944 and 1945. A modest, aging, but significant museum to General Joseph W. Stilwell commemorates his command of all U.S. forces in the China-Burma-India theater from 1942 to 1944. The Three Gorges Dam is expected to bring renewed prominence to the city.

Our delegation's last official stop was Beijing, where we met with a group of professors from the Institute of Geography and Natural Resources of the Chinese Academy of Sciences. Members of this research and graduate-training institute had long been involved with the problems associated with filling the reservoir of the Three Gorges Dam, and the group that met with us gave what appeared to be a frank and realistic appraisal of how the problems were being dealt with. It had been recognized early in the planning stages for the dam that hundreds of thousands of people would be displaced. Part of the discussion centered around exactly what the number was, for some of the delegates had heard numbers as high as two million. According to the researchers, the number at the beginning of the project was about eight hundred thousand. Higher numbers resulting from such factors as the growth of population were expected to bring the actual number when the dam is completed and the reservoir is full to about 1.3 million, we were told.

The researchers seemed to appreciate the plight of these people but stated that the greater good for the larger Chinese population justified going ahead with the dam. This was the bottom-line conclusion of the admittedly small sample of people we met and spoke with, formally and informally, in China.

We spent several days in Beijing and its environs, which enabled us to visit Tian'amen Square, the Forbidden City, the Great Wall, and other tourist attractions. The scale of each of these is enormous, like that of the Three Gorges Dam. Perhaps it is in the nature of a country with a population estimated at 1.3 billion to build on a massive scale, even if in the process one in a thousand of its citizens is displaced. Great projects have great consequences—good and bad—no matter where they are built.

Fuel Cells

In his 2003 State of the Union Address, President George W. Bush called for promoting energy independence for the United States, while at the same time making dramatic improvements in the environment. The familiar rhetoric alluded to a comprehensive plan involving efficiency and conservation, as well as developing cleaner technologies for domestic energy production. But the president soon departed from the familiar and entered the realm of the exotic when he asked Congress to take "a crucial step, and protect our environment" in distinctly new ways. In the twenty-first century, he continued, "the greatest environmental progress will come about, not through endless lawsuits or command and control legislation, but through technology and innovation." He proposed spending $1.2 billion on research on hydrogen-powered automobiles, which employ fuel cells.

The president went on to give an admirably concise definition of the principle of a fuel cell: "A simple chemical reaction between hydrogen and oxygen generates energy, which can be used to power a car producing only water, not exhaust fumes." He challenged scientists and engineers to overcome obstacles to taking fuel-cell-powered automobiles "from laboratory to showroom" in a time frame expressed not in cold calendar years but in the very warm and human image of growing up, "so that the first car driven by a child born today could be powered by hydrogen." Such progress, the president asserted, opened up ways to protect the environment "that generations before us could not have imagined."

It was only in the summer before President Bush's address that I was introduced to fuel cells in a more than passing way. Evidently because I had written about invention and the evolution of a wide variety of technologies, I was invited to join an industry advisory committee that was being formed by Chrysalix Energy, a Vancouver-based venture-capital firm investing in early-stage fuel-cell technology, among the founding partners of which is Ballard Power Systems. It was as a member of that committee, the purpose of which was to bring outside perspectives to discussions about an imagined hydrogen economy, that I received my introduction to fuel cells and became increasingly interested in the history, status, and future of the technology.

Applications of the fuel cell may seem futuristic, but the device itself dates from 1839, when the Welsh-born British jurist and scientist Sir William Robert Groves devised a "gas battery." Unlike Alessandro Volta's invention in 1800 of the now-familiar dry cell, the energy-producing ingredients of which are all contained within the battery casing and can produce electricity only as long as they can sustain the chemical reaction, Groves's gas battery produced electricity as long as it was fueled by an external source. In one modern version of the fuel cell, cathodes are separated by a thin electrolytic membrane. When hydrogen is introduced under pressure, it is decomposed by a catalyst at the anode into electrons and protons. The electrons naturally make electricity, which can power a motor or other electrical device. The protons move through the membrane toward the cathode, where another catalyst recombines them with spent electrons and oxygen from the ambient air to produce water. But even before the fuel cell was invented, the simpler voltaic pile had been developed as a source of electricity, and so the more complicated device became shelved for a century or so, during which time great strides were made in motive power.

Within a couple of decades of the invention of the dry-cell battery, Michael Faraday demonstrated the principle of the electric motor and soon thereafter that of electromagnetic induction, which led to the electric generator. By the early 1830s, working electric motors were being made, and well before the end of the decade electric-driven road vehicles and paddleboats were the subjects of experiments. By 1859, an early version of the lead-acid battery that is used in today's automobiles had been developed, with the most important property of being able to be repeatedly discharged and recharged. As early as 1873, storage batteries were powering electric motors and driving vehicles, and by 1882 they could reach speeds of almost ten miles per hour and travel distances as great as twenty-five miles. The first demonstration of a vehicle powered by an internal-combustion engine was still a couple of years away.

At the end of the nineteenth century, an electric vehicle held the world speed record of sixty-one miles per hour, and in the United States in 1900 almost as many electric-driven cars (1,575) were being manufactured as steam-driven ones (1,684). Combined, they outnumbered by more than three to one gasoline-engined cars. The electric vehicle had the clear advantage of quietness over its then-unmuffled competitors, and it did not need to be hand-cranked to be started. However, after the introduction of Henry Ford's Model K in 1906 and his Model T shortly thereafter, the internal-combustion engine became the power source of choice. By 1912, there were nine hundred thousand gasoline-powered vehicles in America, outnumbering the electrics thirty to one. At about the same time, the self-starter and muffler were introduced, thus making the internal-combustion engine more user-friendly and desirable. It was so much more convenient to add gasoline to a tank than to recharge heavy batteries that did not give a car as much range as gas, even if extra canisters of the fuel had to be carried on a drive of any distance. (Different energy sources are now usually compared by a measure known as energy density, which is the ratio of power to weight. Today, a conventional lead-acid battery has an energy density of about thirty-five watt-hours per kilogram compared to gasoline's two thousand. Although more exotic types of batteries have higher energy densities than lead-acid, they remain an order of magnitude smaller than gasoline and are more expensive to manufacture.) The use of gasoline-powered vehicles in World War I conditioned a lot of young veterans to favor the internal-combustion engine. The last new model of electric car to be built in America during that era was introduced in 1921—at a price four times that of a Model T. The electric vehicle essentially went into forced hibernation for decades, until environmental and energy crises reawakened interest in a nonpolluting alternative to the internal-combustion engine.

Unfortunately, battery technology had not advanced sufficiently in the meantime to enable electric cars to be made attractive competitors to gasoline-driven ones in terms of size, range, and cost. Prompted in part by the rise in consciousness over ecological issues, electric vehicles began to appear again around 1960, but only in small numbers. Indeed, it was not until 1990, when a southern California regional agency—the South Coast Air Quality Management District—required that by 1998 large manufacturers have 2 percent of their sales be zero-emission vehicles, with escalating percentages in subsequent years, that major automobile companies began to look more seriously at alternatives to the internal-combustion engine. And among the alternatives were fuel cells, for which in the meantime suitable electrolytes had been developed for specialized applications. In particular, the Gemini and Apollo space programs employed fuel cells to provide power, while at the same time providing potable water as a by-product. It was the explosion of an oxygen tank and the consequent damage to a pair of fuel cells that produced the life-threatening situation on the Apollo 13 mission, but it was also fuel cells that served reliably for so many other space flights. It was a matter of matching the technology to the application.

Just as dry-cell batteries come in a large variety of types and employ a wide range of different materials—lead, nickel, cadmium, sodium, lithium, aluminum, zinc—so do fuel cells. The ones powering spacecraft typically operate at relatively low temperatures and have an alkaline potassium-hydroxide electrolyte. Small stationary generators operating at intermediate temperatures may use phosphoric acid. Some fuel cells, which are suitable for large stationary generators, operate at high temperature and use solid oxides or molten carbonate as electrolytes. One of the most promising configurations has proved to be the proton-exchange membrane cell, which operates in the temperature range associated with internal-combustion engines and has a power density that makes it suitable for automobiles.

The wide variety of fuel cells under development is to be expected in a fledgling industry populated by hundreds of private and scores of public companies. Among the best known of these is Ballard Power Systems, which is most closely associated with the proton-exchange membrane technology. After receiving a degree in geological engineering from Queens University, Canadian-born Geoffrey Ballard began his career in oil exploration, but he became increasingly frustrated when his opinions were ignored in favor of those from scientists with higher degrees. After going back to school and earning a doctorate in geophysics, Ballard began working for the U.S. Army as a civilian. In this position, he was exposed to management methods and got to know his way around the military-industrial complex. During the 1973 energy crisis, Ballard was made director of research for a new government office of energy conservation. He was excited about the position and the possibilities, but he became disillusioned when he found that the research-and-development funding system expected results much more quickly than the twenty-year gestation period that he knew would be required for new energy systems.

Ballard also became convinced that conservation was not the answer to the energy problems facing the United States and, especially, the Third World, whose people wanted to enjoy the level of abundance they saw in America and did not want to have to conserve to get there, if indeed that was a possible route. The real need, Ballard believed, was to develop energy-conversion devices and techniques that were more efficient and cleaner than traditional fossil fuels. He also saw a need for lightweight and compact portable power sources. Such miniaturization would lead to developments that would revive interest in electric vehicles. Thus, Ballard left his job in energy conservation and struck out on his own to develop smaller, lighter, and more efficient batteries to power everything from video cameras to light trucks and delivery vans. He and a financial backer enlisted a chemist, who went to work developing a lithium battery, which in turn attracted further support to power a submarine. The battery work involved living from hand to mouth, and so the Ballard team was always looking for new funding opportunities.

In the meantime, the company, which had started in Arizona, had relocated to Vancouver, and so was on the lookout for potential Canadian government contracts. One request for proposals that appeared promising was for the development of a low-cost solid-polymer fuel cell. Since this involved electrochemistry, something the Ballard group had become heavily involved with in its battery work, the project seemed like an ideal one to bid upon. Ballard had had some experience with fuel cells while working for the U.S. Army, and he knew that the technology was working in the space program.

The challenge was not to demonstrate that fuel cells worked but to demonstrate that they could be produced for the consumer market. This meant, of course, that they had to be manufactured with a much higher power density and at a much lower cost. With the help of venture capital, Ballard went on to produce stationary fuel cells, but the real challenge lay in demonstrating a fuel-cell-powered vehicle, which meant fitting the power plant—and the hydrogen supply to fuel it—into a space with predetermined practical requirements and limitations, such as a bus.

In the meantime, in order to secure more capital, the founding members of Ballard had given up control of their company to a management team, which at first did not like the idea of a bus project. But after Geoffrey Ballard secured federal and provincial support for the bus idea, the company also bought into it. The prototype vehicle proved to be a resounding success when it had its public rollout in 1993.

The rollout turned out to be literally that. As the bus stood idling beside Vancouver's Science World, ready to be driven around to the front as soon as the speeches were finished, the compressor suddenly quit. The problem was a broken bolt, and there was no time to fix it. So some Ballard employees surreptitiously pushed the bus to get it started moving, after which it rolled down an incline and was steered to a stop in front of the podium. Since fuel cells were expected to be quiet, no one noticed that the bus had not been under power. After Ballard, who had been alerted to the crisis, and others had spoken of the significance of the occasion, the British Columbia premier announced, "Let's go for a ride." Fortunately, a large crowd had gathered around the bus, and reporters began to question the man responsible for the bus project, Paul Howard. He answered all their questions with great patience, finally telling everyone to come back that afternoon for a ride. The broken bolt was fixed during lunchtime, and the bus ran smoothly for the news cameras. Today, later-generation fuel-cell buses are carrying passengers in Madrid and other European cities.

Geoffrey Ballard is no longer with the company that bears his name, but the fuel cell that he was so instrumental in promoting to "change the world" has reached an unprecedented level of respect as an alternative energy system. Ballard Power Systems, which has registered the trademark "Power to Change the World," is forging full speed ahead, having established partnerships with DaimlerChrysler, Ford, and other automobile manufacturers. The goal is not only to supply the transportation industry with fuel-cell engines but also to develop fuel-cell systems for stationary equipment and portable devices. And Ballard is, of course, not the only player in the game.

In the meantime, recognizing that production-line fuel-cell vehicles are still a ways off, automakers began promoting hybrid cars in the late 1990s. Toyota was the first to market. Its Prius, introduced in Japan in 1997, had cumulative sales of one hundred thousand worldwide within five years. Honda followed with its two-seater Insight, and soon thereafter offered a hybrid version of its Civic. Hybrids are primarily gasoline-powered, but they employ an electric motor for use at low speeds and to assist in high-acceleration driving, thus conserving fuel. (Rated gas mileage can be of the order of fifty miles per gallon.) By having their batteries charged by the onboard gasoline engine, hybrids overcome the disadvantages of pure electric vehicles. Also, since the engine runs at a low and fairly constant speed, high efficiency is matched by low maintenance costs. DaimlerChrysler is expected to offer a hybrid pickup truck soon, General Motors has promised to have hybrid power in five vehicle models by 2007, and Toyota has announced plans to sell worldwide as many as three hundred thousand hybrids annually by then. But hybrids are still essentially gasoline-powered, with the internal-combustion engine being their ultimate source of electricity.

The hybrids thus do not fully address environmental concerns, and they are seen by some observers as transition vehicles that will ease the radical change from the petroleum-based economy to the hydrogen-based one assumed to be required for the full-scale adoption of fuel-cell-powered vehicles, which are currently still in the concept phase. Unfortunately, at least in the early stage of fuel-cell use, hydrogen is likely to be produced from a hydrocarbon-like natural gas, which is a less efficient process than that used to convert oil into gasoline for use in a conventional automobile.

The General Motors concept car termed Hy-wire is not only powered by a fuel cell but also controlled through "by-wire" technology similar to that already widespread in the aircraft industry. There are no mechanical linkages between driver and throttle, steering, or brakes, since all such connections are by electrical wire, which leaves room for a more imaginative overall vehicle design. Since there do not have to be mechanical linkages between pedals and steering wheel and what they normally control, there do not have to be pedals or a steering wheel at all. Hence, the Hy-wire (a portmanteau word formed from "hydrogen" and "by-wire") vehicle is often described as a "skateboard design," in which the fuel cell and appurtenances are incorporated into a rather flat chassis, onto which a variety of body types can be mounted (and changed like clothing to fit the mood of the owner). Because there are no mechanical linkages between body and chassis, the imagination of automobile designers is freed up to reconfigure the interior, which can mean a roof-to-bumper windshield and a handheld control system that is not unlike that of the video-game controllers with which younger generations have grown up. The concept car is also referred to as the China car, since it is expected to become available as early as 2008 but no later than 2015, or simultaneously with what has been perceived to be a potentially booming auto market in Asian countries and elsewhere around the world, in which private vehicle ownership currently rests at 12 percent.

Fuel cells are expected not only to revolutionize the appearance and control of automobiles but also to greatly change the perception of them as noise and air polluters. Since the fuel cell itself has no moving parts, the only sound associated with its operation is that of the delivery devices needed to supply the fuel, but reportedly not everyone likes the sound of the compressor in the Ford system. (A significant lack of engine noise was the most striking feature of the first electric vehicle I rode in.) Also, since the only by-product is water vapor, in place of smelly exhaust fumes, there will be but wisps of warm vapor or drips of distilled water. Instead of the present image of internal-combustion vehicles as having developed into greater polluters of cities than the horses that they began to displace about a century ago, fuel-cell-powered vehicles have the potential for being seen as saviors of the planet—if the problem of generating hydrogen in an acceptable way can be solved.

But fuel-cell-driven vehicles can be successful only if there is an infrastructure in place whereby they can be refueled. The "hydrogen economy" will become a reality only when the elemental gas is as readily available as gasoline is now. Though hydrogen is the most abundant element in the universe, its gaseous form does not occur naturally on earth. Thus, another source of energy must be employed to produce free hydrogen from its compounds. Among the sources of hydrogen are natural gas and other hydrocarbon fuels, but these are nonrenewable resources, and the greenhouse gas carbon dioxide is produced in the process of releasing the hydrogen. Hydrogen can also be made from water, by electrolysis, but it is not as efficient a process. Thus, how to extract large quantities of hydrogen cleanly and efficiently remains a topic of some debate.

In the United States currently, hydrogen-powered cars and buses must be refilled at specialized sites, but Shell Hydrogen has announced plans to have a dispensing system for the new fuel integrated into a regular gas station in the Washington, D.C., area shortly, so that hydrogen can be obtained much the way diesel fuel is today. The cost of outfitting just one such station with the necessary tanks and pumps has been estimated by BP America to be between five hundred thousand and one million dollars. At the higher estimate, making hydrogen available at thirty thousand gas stations would involve a thirty-billion-dollar investment, but this is not considered out of the question for introducing a new energy source. A pipeline to bring natural gas down from Alaska has been priced at more than twenty billion and a single nuclear plant can cost more than ten billion.

Thus, the widespread availability of hydrogen at corner locations, although it requires a major commitment on the part of fuel distributors, is in fact within reach. However, the hydrogen economy still faces the familiar catch-22 associated with technological change: The infrastructure needed to facilitate a paradigm shift is not likely to be put into place until the paradigm shifts, but that is not likely to happen until the proper infrastructure exists. In the final analysis, change depends upon technological pioneers who are willing to undertake rough rides over unpaved roads and carry their toolboxes and extra fuel along with them. There are other obstacles on the road to a hydrogen economy, in which not only vehicles but also everything from large electric generating plants to handheld electronic devices are fitted with fuel cells and so powered by gaseous hydrogen, on which they run most efficiently. (Using high-temperature fuel cells allows natural gas to be used directly, which significantly lessens the problems associated with establishing a fuel-delivery infrastructure.)

By the time fuel-celled vehicles are on the road in large numbers, they can be expected to carry enough hydrogen to give them a range as great as today's gasoline-powered vehicles. Early demonstration vehicles powered by fuel cells had to be fitted with large hydrogen-storage tanks, which either encroached on interior space or perched obtrusively on the roof. This is still not much of a problem for buses, but it certainly is for stylish automobiles. High-pressure tanks capable of containing hydrogen at ten thousand pounds per square inch have reduced the storage problem somewhat, but it is still something that has to be dealt with before sleek, roomy vehicles with competitive ranges hit the show-room floor.

Since it is not likely that we would want to have to carry hydrogen tanks on our backs to power cell phones or laptops, fuel-cell technology is also developing on a smaller scale. In fact, some industry followers believe that the first big inroad of fuel cells into the marketplace will be in the area of consumer electronics, in which customers have already demonstrated a willingness to pay a premium for novelty and convenience. Indeed, it has been said that "half the interest in fuel cells is out of frustration with batteries." Fuel cells can provide power for a much greater amount of time than batteries, and thus they are expected to free laptop users and television-camera crews from having to lug around heavy battery packs. One benchmark of the portable-power industry is how many hours of continuous power can be gotten out of a kilogram of fuel. The comparison is striking: pure hydrogen, as high as 38,000 hours; methanol, from which hydrogen can be extracted, 6,000 hours; a fully charged lithium-ion battery, 150 hours. (But these numbers do not include the size of the container, which can be much larger for a kilogram of hydrogen compared to that of methanol.) Unfortunately, unless they are carefully matched to the application, fuel cells are not capable of delivering large bursts of power. Hence, it is foreseen that they will be combined with batteries in many applications, exploiting (like hybrid cars) the advantages of each. And the same distribution system that currently makes batteries available at every discount outlet, supermarket, and convenience store can provide fuel cartridges at those locations, too.

Recent developments in the consumer-electronics industry are likely to accelerate the move toward fuel cells as power sources. Sony, Samsung, and others have begun introducing products that combine previously separate devices, such as a cell phone, PDA, digital camera, and MP3 player, with the capability of being continuously connected to the Internet. Such multifaceted electronic gadgets also come with relatively large color screens, and their power requirements will thus consume batteries as quickly as teenagers do soft drinks. Electronics manufacturers are hence experimenting with powering beta versions of the multi-devices with fuel cells.

There is also considerable effort going on in other areas. In Japan, a Ballard partnership recently unveiled a precommercial version of a stationary one-kilowatt fuel-cell generator that is capable of providing private residences with power and heat. The system was expected to be available in the marketplace by the end of 2004. Ballard, which has stated that it wants to complete its transformation "from a technology-focused research and development organization into a customer-focused production organization" appears to be well on its way to fulfilling its objective.

Ballard is not the only player moving ahead in the fuel-cell business. For a couple of years now, six two-hundred-kilowatt fuel cells have been providing power for a juvenile training school in Middletown, Connecticut, a state that has encouraged clean energy and that is home to a number of small start-up companies focusing on fuel-cell research and technology development. Connecticut is also the location of fuel-cell manufacturers, including UTC Fuel Cells in South Windsor, which at the end of 2002 was said to be the "world leader in fuel cell production at this point in time." The fuel-cell power plant at Middletown not only provides electric power for the campus but also is used to heat and cool the buildings. It came at a high price, however, for the installation of the forward-looking power plant accounted for about 37 percent of the $49 million total cost of the 227,000-square-foot facility.

The decision to employ fuel cells in Connecticut might not have been made without government involvement, which promises to increase.

President Bush's enthusiasm for a hydrogen future brought considerable renewed attention to what has for so long been an obscure technology promising clean and quiet power. Given such high-level support, and assuming an acceptable means of producing hydrogen will be developed, it is a good bet that before long today's exotic technology will be familiar, not only in North America but around the world.

Engineers' Dreams

On numerous occasions after writing or lecturing about large engineering projects, I have been asked if I know of a particular book published in the 1950s. Few of the questioners can remember the book's exact title or the name of its author, but all have a vivid recollection that the book described grand schemes of engineers, ideas such as a tunnel between England and France, a dam across the Strait of Gibraltar, and widespread use of solar and tidal power. The book I believe all my questioners have in mind is one by Willy Ley, published in 1954. Coming across it can be as exciting an experience for the first-time reader today as it was for those who remember discovering the book as young men and women five decades ago. Both the book and its author, and the contemporary influences on them, deserve to be remembered.

Willy Ley was born in Berlin, Germany, in 1906. He attended public schools and developed a broad interest in things scientific and technical. During the poor economic times of the 1920s, Ley attended on and off the universities of Berlin and Königsberg, where he took courses in paleontology, astronomy, and physics. He had planned a career in geology, but in 1926 he encountered an early book on space travel, one written by Hermann Oberth, who himself had been drawn to the subject by reading Jules Verne novels.

Oberth was born in Transylvania in 1894, when it was still part of the Austro-Hungarian empire. After his studies as a medical student at the

University of Munich were interrupted by World War I, he became increasingly interested in other matters and, in time, submitted a doctoral dissertation on a rocket of his own devising to the University of Heidelberg. His thesis was rejected by the university, however, as it was subsequently by several book publishers. In 1923, Oberth, who is now considered one of the founders of spaceflight, did publish *Rocket into Interplanetary Space*, complete with equations and technical analysis. Among his other endeavors, Oberth soon convinced the film director Fritz Lang to back what would be a failed attempt to launch a rocket to publicize the movie *Woman in the Moon*. After a peripatetic career, which included work with Wernher von Braun at Peenemünde as well as at the Redstone Arsenal in Huntsville, Alabama, an aging Oberth began to disconcert his admirers by speculating about unidentified flying objects and supporting parapsychology.

After reading Oberth's 1923 book on rocketry, Willy Ley abandoned his own formal education to follow a growing popular preoccupation in Germany: thinking about travel beyond the stratosphere. Ley's first book, *Trip into Space*, a popularized version of Oberth's work, was published in Germany in 1926. Isaac Asimov wrote that Ley's book "met with wide acclaim and for over forty years afterward he remained the most successful popular writer on rocketry in the world."

In 1927, Ley became one of the founders of the German Society for Space Travel, serving as vice president from 1928 until 1933, when the society was dissolved by the Nazis, who did not want information on rocketry widely disseminated. Ley's linguistic skills enabled him to correspond with others about rocketry throughout Europe and America, and his society became a clearinghouse for information on the subject. He introduced a young Wernher von Braun into the rocket society; decades later, the two would coauthor a book on the exploration of Mars. Ley, who has been described as von Braun's "first tutor in rocket research," also wrote with him about the conquest of the moon several years before *Sputnik*'s launch prompted President Kennedy's vow to put a man on the moon. According to Asimov, Ley, "more than anyone else, prepared the climate within the United States for the space effort."

But before Ley—or von Braun for that matter—would get to America, he had to deal with the increasing interest of the Nazis in rocketry. By 1935, Ley had been ordered to stop writing about the topic in foreign publications, which he did, but some of his articles submitted before the ban were continuing to appear. He sought help from British and Dutch friends and soon left Germany for an "extended vacation" in England. An invitation from the American Rocket Society, then known as the American Interplanetary Society, soon took him to the United States. Ley found little interest in rocket theory among Americans generally, and so he began to write on zoology and other scientific subjects for popular magazines. In 1940, he joined the New York newspaper *PM* as science editor, and the following year he published his first books in English, including *The Lungfish and the Unicorn: An Excursion into Romantic Zoology,* which was a popular success, and *Bombs and Bombing: What Every Civilian Should Know.* With interest in rockets growing during the war, Ley returned to his first love and in 1944 published *Rockets: The Future of Travel Beyond the Stratosphere,* a book that dealt with German and Russian developments and attracted a wider readership after German V-2 rockets began to rain down on London. That same year, Ley left journalism to join the Burke Aircraft Corporation in Atlanta as director of engineering, testimony that his insights into rocketry were technical as well as journalistic. Burke Aircraft soon became part of the Washington Institute of Technology, and Ley was based in College Park, Maryland, near Washington, D.C. His work there earned him a listing in the 1948 edition of *Who's Who in Engineering,* in which he was identified as a research engineer and rocket expert. (Among his other contributions, he is credited with "devising the basic principles that led to the development of the liquid-fuel rocket," although some of this credit might have been due Oberth.)

Ley continued to maintain an interest in science fiction as well as in rocketry, and when he left the Washington Institute in 1948, he began to write increasingly about space travel. His book *The Conquest of Space,* which appeared in 1949, received the nonfiction prize at the First International Fantasy Fiction Awards. In the 1950s, at the Hayden Planetarium in New York, Ley gave public lectures on space travel, and he served as a consultant to television and movie projects dealing with the subject. In one movie on which he worked, *Woman in the Moon,* the same Fritz Lang film with which Oberth was associated, the rocket-launching countdown from ten to zero is said to have been introduced. Although Ley continued to write on natural science and other topics, he concentrated on space travel, contributing to many books on the subject. His book with von Braun, *The Exploration of Mars,* appeared in 1956; *Beyond the Solar System,* with a foreword by von Braun, appeared in 1964; and *Ranger to the Moon* came out in 1965. Ley also wrote *Watchers of the Skies: An Informal History of Astronomy from Babylon to the Space Age,* which was published in 1963. In an article in *Science Digest* in 1951, he predicted that "the year of the first spaceship probably will be 1965 or 1970." He once confessed that his foremost wish was to "live to rise to the moon," but Willy Ley died just three weeks before the first moon landing in 1969.

With the conquest of the moon and other accomplishments in space, Ley's books on such subjects, no matter how captivating when they were first published, could not be expected to command the same attention of space-age readers, and with the advancement of space technology they remain now largely of historical interest. However, a book Ley published in 1954 has retained the power to captivate the reader, because in this book he dealt with terrestrial things that by and large have not yet been achieved. The book, *Engineers' Dreams,* is the one that so many people remember, and its continued currency stems from the fact that it deals with proposals for earthbound projects of such magnitude and potential environmental impact that even when couched in the technology of a half century ago, they are awesome reminders of how even the most mundane engineering, when carried out on a grand scale, can reshape our planet.

As Ley writes in the early pages of *Engineers' Dreams,* "the word fantastic, when applied to engineering, merely means 'it has not yet been done.' " Some of the once fantastic things that are described in the book have been worked on, albeit with varying degrees of success, and in at least one case fully achieved, in the past five decades. These include such projects as solar power, wind and wave energy, and the Channel Tunnel, which have indeed ceased to be fantastic. Another postwar dream that Ley articulated, that of floating airports, was in effect realized in the late-twentieth-century artificial-island airports in Hong Kong and Japan.

However, none of the projects Ley described was much closer to reality a decade after *Engineers' Dreams* first appeared, for the nine chapters in the original edition needed hardly any updating when a revised version appeared in 1964. The book was expanded to include two new chapters, one dealing with ongoing Dutch efforts at reclaiming land from the North Sea and one dealing with Russian efforts to stabilize the Caspian Sea. It is such map-altering projects that form the core of the first edition of *Engineers' Dreams,* and it seems hardly surprising that in revising it Ley added more of them to a book whose staying power seems to have been ensured by its descriptions of some of the oldest engineering techniques applied on some of the grandest scales ever. Among the still fantastic projects are the creation of two large lakes in central Africa and the reduction of the size of the Mediterranean Sea, each to be accomplished by the building of dams.

As he does with other grand proposals, Ley provides some historical perspective before describing the Africa project. He relates how, when much of that continent was still being explored in the nineteenth century, the problem of transportation was widely discussed. The problem was especially acute where there was not a convenient river, and thus the worst situations of all arose in riverless northwestern Africa, where the Sahara Desert is located. Since the Sahara is enclosed by hills and mountains, explorers approaching it from all sides had the distinct impression that it was well below surrounding land. Indeed, they thought it might be below sea level; if so, it could be flooded by digging a canal through the hills between the desert and the Atlantic Ocean. With the Sahara flooded, oceangoing ships could sail into a Saharan sea, and travelers could easily reach Morocco from the south and Nigeria from the north and possibly sail all the way across the continent to Egypt. It was only when further exploration revealed that in fact only very few areas of the generally above-sea-level Sahara could be flooded by such a scheme that it had to be abandoned.

All this is by way of background to the African project that is the focus of one of Ley's chapters. It involves Lake Chad, which he describes in the mid-century as having shrunk, mostly because of evaporation, to a mere 6,000 square miles from the 170,000 it covered 10,000 years ago. This dearth of water occurs only a few hundred miles north of rain forests where water abounds. Through the rain forest also passes the Congo River, which is believed to have originated when the water from a vast inland lake—at the location of the present Congo Basin—flowed through the mountains to the sea. Now the Congo passes along chasms that were eroded through the mountains, and building a dam in one of these would back the water of the river and its tributaries up into the basin to re-form a Congo lake, the 350,000 square miles of which would nearly equal the combined area of all of California, Nevada, and Oregon. Some overflow from the lake could be fed northward into the Chari River, which is the main source of water for Lake Chad, thus restoring conditions that fed the Chad Sea ten thousand years ago. In time, the restored Chad Sea would also begin to overflow, and the excess water could be directed into a newly navigable river to flow ultimately into the Mediterranean Sea.

Today, of course, there would be considerable discussion of the environmental impact of such an enormous and far-reaching project, but at the time Ley was writing, there was not yet a general sensitivity to such issues. He describes some of the benefits of the project, such as the availability of huge amounts of hydroelectric power and new means of water transportation. He acknowledges that such massive flooding could destroy some places of scenic beauty, but he generally believes that "everything of known value" would be outside the area to be inundated. Ley does recognize that, even though some of the land to be flooded is a breeding ground for tropical diseases, there would be a large number of people living there who would have to be relocated. He rationalizes, however, that since their living conditions would be improved "it is unlikely that they would object." He does concede, on the other hand, that changing African political boundaries could make the project unrealizable. Nevertheless, on balance, Ley's recounting of a cost-benefit analysis betrays the insensitivity of his times to issues that today evoke virtual knee-jerk reactions, even when the projects proposed are minuscule and innocuous in comparison to damming the Congo River.

The Congo plan was in fact the brainchild of Herman Sorgel, an architect by training who came to be employed by the Bavarian government. Sorgel, who was born in 1885, had written on architecture and aesthetics, including books on Frank Lloyd Wright and the weekend house, before turning his attention to things on a larger scale, which related to his growing interest in geopolitics. He envisioned three world supercontinents, one of which would be formed by uniting Europe and Africa, with the latter providing resources and living space for the growth of the former. His concern for the peoples of Africa seems to have been virtually nonexistent.

If Europe and Africa were to be united, there would have to be land bridges and vast sources of power, and thus, in the late 1920s, Sorgel outlined his plan for reclaiming land from the Mediterranean Sea. At first he called it the Panropa Plan, but since there was a similarly named political group promoting European confederation, Sorgel renamed his scheme the Atlantropa Plan. He first published his idea early in 1928, followed the next year by a more comprehensive plan entitled "Reduction of the Mediterranean."

There are four main sources of water that replenish what the Mediterranean loses to evaporation: rainfall, the Black Sea, rivers, and the Atlantic Ocean. The flow through the Strait of Gibraltar accounts for two thirds of the total inflow, and so erecting a dam across the strait would cause the level of the Mediterranean to fall about forty inches per year over its one-million-square-mile expanse. In order to be positioned across the strait so that its height nowhere exceeded about one thousand feet, such a dam might have to be on the order of twenty miles in length. Once a dam was in place, in a decade's time the level of the Mediterranean would drop more than thirty feet, and at this difference in elevation the water from the Atlantic could be sent through turbines to generate an enormous amount of electricity. After a century, the 330-foot drop in water level would uncover ninety thousand square miles of new land, resulting in the joining of the islands of Majorca and Minorca and of Corsica and Sardinia.

Because of the undersea topography, the western part of the Mediterranean would not be expected to change much beyond one hundred years after the completion of a Gibraltar dam, and Sorgel proposed that at that time two more dams be constructed, one between Italy and Sicily and one between Sicily and Tunisia. With the western Mediterranean enclosed, its water level could be maintained at 330 feet below present sea level. The eastern Mediterranean could be allowed to continue with a net loss of water for another century. Although the Mediterranean area would benefit by about 220,000 square miles of new land and vir-

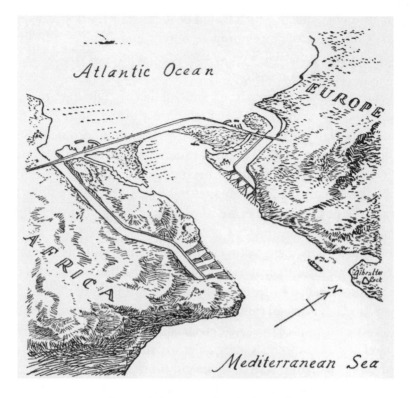

Proposed dam across Strait of Gibraltar

tually unlimited quantities of hydroelectric power, the lost water from the entire Mediterranean Sea would amount to adding possibly as much as three feet to the level of the oceans worldwide.

The recognition that such a drastic rearrangement of the world's waters would be opposed by many caused Sorgel to emphasize that the drop in the Mediterranean could be stopped at fifty feet. This would mean that only a single-lock canal might suffice for getting oceangoing ships into and out of the Mediterranean and would preserve the value of most existing harbors. Dropping the sea level as much as one hundred feet would require additional locks at Gibraltar and at the Suez Canal, and today's principal Mediterranean harbors would be far from the sea and thus useless. Even supporters of Sorgel's plan, who saw its power-generating aspect as a distinct benefit, argued against too much of a lowering of the Mediterranean's water level, because the lessened weight of water on unstable areas of the seafloor might lead to earth-quakes or volcanic eruptions.

Only in Willy Ley's closing remarks on Sorgel's Atlantropa Plan does he recognize that there are enormous technical challenges to building a dam across the Strait of Gibraltar. However, Ley found it "unnecessary to discuss the engineering difficulties" because the political situation of the 1950s made "the engineering problems future problems, and we can't tell how an engineer 50 years from now would go about solving them." It has now been fifty years since Ley wrote those words, and it is clear that such a project would not be given anywhere near the serious consideration it was when Sorgel first proposed it—in a distinctly different political climate. The sensitivity of engineers and politicians alike to environmental impact and the sovereignty of states and peoples make a Gibraltar or Congo dam more than a technological improbability in the foreseeable future.

Nevertheless, engineers continue to dream, and among their dreams that were not but could have been included in Willy Ley's *Engineers' Dreams* are, as we have seen, bridges across the straits of Gibraltar and Messina. For all of the technical challenges, such fixed crossings are minuscule projects, with minuscule environmental and political implications, compared to the dams envisioned by Ley, and they are likely to be realized in the present century. Other projects that are being promoted today as seriously as those Ley wrote about a half century ago include the fixed link between Alaska and Siberia, thus making world-wide train travel a reality. Another great project envisions the damming of Hudson Bay, thereby providing a vast new supply of freshwater for North America. The engineers and dreamers of such macroprojects will no doubt continue to promote them. As Frank Davidson hopes at the close of his 1983 book on macroengineering projects, *Macro,* a work that might be seen as a successor to *Engineers' Dreams,* the discussion of joint technological undertakings of unprecedented magnitude may well someday be added to the agendas of summit meetings as alternatives to war and warfare.

Engineers' Achievements

W e mark time differently than we measure distance. Although few of us are likely ever to have seen the odometer on a new car read 000000, we infer that that was indeed the reading when it was installed at the factory. As the car was driven from the assembly line to its parking lot, the wheels began to turn, and the machine registered tenths, miles, tens of miles, and so forth, as they were reached. With dates we mark the days, months, and years before they are completed, however, and so from the instant after midnight on New Year's Eve 1999, we by custom wrote the date as 1/1/00, even though it would be another twenty-four hours before that day was completed. And it was not until midnight on 12/31/00 that the year 2000 was completed, and with it the twentieth century. The imagined computer nightmare labeled Y2K notwithstanding, the second millennium ended not with a bang but a whimper.

Many people knew all this, of course, but it would have been difficult for them to hold out for an entire year before recognizing the arrival of the twenty-first century and the "new millennium," which truly began the instant after midnight on 12/31/00—that is, on 01/01/01. If we had insisted on waiting until 2000 turned into 2001 for our celebrations, we would have risked being seen as spoilsports at best or hopelessly out-of-date at worst. Ironically, it was the focus on the very technical problem of Y2K computer compliance, a situation closely associated with engineers and scientists, or at least computer scientists, that emphasized the four-digit year, which in turn drove the focus of the date change to the odometer-like event of 1999 turning into 2000. The year change was treated as a counting event, when in fact it was a marking event.

Not since the upside-down year 1961 almost four decades earlier or the upside-down, palindromic, and mirror-image year 1881 more than a century ago had a new year's number been treated as such an icon. Such oddities will not occur again until the numerologically rich seventh millennium gives us the year 6119 and the ninth millennium gives us 8118. In terms of numerical patterns, 1999 and 2000 are identical—a single digit followed by another digit repeated three times. However, the freshness of the preface numeral 2, after one thousand years beginning with the numeral 1, seemed special and remarkable, especially when followed by three zeroes. The change in the prefix attracted the kind of attention that a decadal birthday or anniversary does, with just about everyone forgetting or abandoning the fact that we measure age by different numerical rules than we mark dates. From newspapers and magazines to professional societies and organizations, millennial madness triumphed over reason—with seemingly countless lists modified by top, best, greatest, and other superlatives being voted on, compiled, and disseminated.

Throughout 1999, the *New York Times Sunday Magazine* celebrated the ten centuries leading up to "the biggest birthday any of us will ever live through" with six special issues. One of these dealt with "the best of the millennium" and another with "10 centuries, 25 turning points," thus coupling our fascination with zeroes with our love of fives. *Time* magazine named the "100 Greatest Minds." As the year-end approached, newspapers and magazines of all stripes provided us, along with the usual top stories of the previous year, the events, inventions, people, and so forth, of the century and of the millennium, albeit each declared finished one year early. Anything comparable that happened in 2000 was not to be so commemorated, in the popular press at least, for another century, if not another millennium. Should it have been any great surprise then that engineering and scientific organizations also threw reason to the wind—along with an opportunity to educate the public about how we use numbers to count and mark things—and jumped on the millennial bandwagon?

Lists are always interesting, of course, even to the supernumerate—and even though they realize the arbitrariness of numbers, including "round" numbers ending in 0 or 5. It is only an accident of our base-ten system and the multiple recurrence of 5, we might tell ourselves rationally, that top-ten lists or twenty-fifth anniversaries are special, but we get caught up in the process of compiling lists, perhaps because they appear to have a rationality of their own, a beginning and an end, an ordering, a definiteness, a decisiveness, a significance. The phenomenon is nothing new.

A survey conducted seventy years previously by the American Society for the Promotion of Engineering Education sought to identify "the outstanding engineers of the past twenty-five years; also those who might fairly be considered the greatest engineers of all time." In addition to the still familiar Edison and Ford listed among the outstanding engineers of the first quarter of the twentieth century, there were John F. Stevens and George W. Goethals, who played central roles in the construction of the Panama Canal, completed in 1914. These two men were also identified as among the greatest engineers of all time, but a 1974 article about Stevens in *Civil Engineering* magazine was entitled "The Forgotten Engineer." (Ferdinand de Lesseps, the French entrepreneur who pushed the Central American canal project, as he had earlier the Suez Canal, was named the fifth greatest engineer of all time, even though he was not even an engineer.) Lists are fickle.

Number ten on the list of outstanding engineers of the first quarter of the twentieth century was Ralph Modjeski, the builder of Philadelphia's Benjamin Franklin Bridge, completed for the nation's 150th birthday, in 1926, and thus a name still quite familiar in 1930. Two bridge builders were also included on the list of greatest engineers of all time, but that is not to say that they got any respect. James Buchanan Eads was misidentified as William B. Eads, and a John L. Roebling was listed—presumably the John A. Roebling who had designed the Brooklyn Bridge.

Also included on the list, among Archimedes and Leonardo, were Charles Steinmetz and John Ericsson. Steinmetz, now a name virtually unknown among younger generations, in the early twentieth century was synonymous with engineering. The stooped-over, cigar-smoking character who explained technology over the still infant radio and who ran for public office in New York State, where he worked for General Electric, seemed to be universally recognized as the genius who brought electricity to the masses. His theoretical calculations made long-distance transmission practical, and his production of lightning in the laboratory made theater. As late as the mid-twentieth century, any engineering student might affectionately be called "Steinmetz," but the allusion would be lost by 2000.

Similarly, John Ericsson is now hardly a household name. Yet this builder of the *Monitor* ironclad "revolutionized navigation by his invention of the screw propeller," as stated on the little-known but prominent monument to him that stands in Washington, D.C., to this day, beside the Potomac River, next to the Arlington Memorial Bridge, and a stone's throw from the Lincoln Memorial. Lists do not have the staying power of monuments, however.

Engineering News-Record celebrated its 125th anniversary during 1999 and did so by identifying the 125 most significant innovations that shaped the construction industry during the magazine's lifetime—as well as the industry's 125 most influential people who made contributions since the magazine's founding in 1874. Eads made the list, but Modjeski did not. (John Roebling's son, Washington, and his wife, Emily Warren, were on the list as project managers, but the elder Roebling, who died in 1869 just as construction was beginning on the Brooklyn Bridge, did not qualify.) Should the magazine identify the top 150 people to celebrate its 150th anniversary in 2024, it should not surprise anyone if on second thought, and in the context of a longer history and more candidates, some of the 125 do not make the cut twenty-five years hence. Lists are products of their times.

In the wake of the Centennial of Engineering celebration, which took place in 1952 and coincided with the one hundredth anniversary of the founding of the American Society of Civil Engineers, the country's first national engineering-professional society, the ASCE undertook to identify outstanding American civil-engineering works. The 1955 list of the seven modern civil-engineering wonders of the United States reflected the importance of water supply, treatment, and control in the first half of the twentieth century, with the majority of the seven wonders being in that category. The list was presented in alphabetical order:

Chicago's Sewer Works
Colorado River Aqueduct
Empire State Building
Grand Coulee Dam
Hoover Dam
Panama Canal
San Francisco–Oakland Bay Bridge

This mid-century list was updated in 1994, perhaps a bit prematurely, to identify "the nation's seven most spectacular civil engineering achievements of the 20th century." In the meantime, traffic had become more worrisome than water, and so the water wonders were largely displaced by wonders of transportation. In only five decades, the list of the seven wonders of the United States was almost completely revised:

Golden Gate Bridge
Hoover Dam
Interstate Highway System
Kennedy Space Center
Panama Canal
Trans-Alaska Pipeline
World Trade Center

Each list contained a single bridge, with the first list's San Francisco–Oakland Bay Bridge, which opened in 1936, being displaced by the nearby Golden Gate Bridge, which opened just six months later, in 1937. The mid-century list recognized the greater engineering achievement of the now lesser-known structure, but the late-century list featured the notable but when-constructed state-of-the-art span that had by then become the outstanding tourist symbol of San Francisco. (The Golden Gate had also been passed over for the six years older George Washington Bridge in the design of a stamp commemorating the centennial of American engineering in 1952.) Engineers are as subject as anyone to fashion and short-term memory: Four of the seven wonders on the revised ASCE list were achievements of the second half of the twentieth century. We all tend to forget what has not been recently in the news.

The National Academy of Engineering, which celebrated its twenty-fifth anniversary in 1989, elected to mark the occasion by announcing a ranked list of ten outstanding achievements of the period of its existence. These were judged to be:

1. Moon Landing
2. Application Satellites
3. Microprocessor
4. Computer-Aided Design and Manufacturing
5. CAT Scan
6. Advanced Composite Materials
7. Jumbo Jet
8. Lasers
9. Fiber-Optic Communication
10. Genetically Engineered Products

The list was not greeted with unanimity, especially among civil engineers, who saw as also deserving of recognition such neglected achievements as the cable-stayed bridge, the tubular skyscraper, and the interstate highway system, which, although not exactly invented during the period, were then certainly pushed to new limits. Indeed, it is a characteristic of engineering that, whatever the period, new achievements will overshadow the old. Something different in kind can always steal the spotlight from something different in scale.

In a booklet promoting the academy's outstanding engineering achievements, the twenty-five-year period ending in 1989 was described as one that "witnessed more advancement in technology and, consequently, greater change in the lives of people than any previous 25-year period in recorded history." This sentiment echoed that expressed ninety years earlier in the *New York Times*, which on December 31, 1899, carried a story on the nineteenth century. As quoted by Stephen Jay Gould in his *Questioning the Millennium*, the story asserted that "tomorrow we enter upon the last year of a century that is marked by greater progress in all that pertains to material well-being and enlightenment of mankind than all the previous history of the race." We can only speculate on what will be written of the twenty-first century.

To celebrate not an anniversary of its own but the arrival of the year

2000, the National Academy of Engineering embarked upon a project designated Greatest Engineering Achievements of the Twentieth Century. The idea, which grew out of a desire to convey the importance and excitement of engineering to the public, especially to young students, was to focus on "the significant impact that engineers and engineering have had on the quality of life in the 20th century." To reduce the chance of omitting—or at least failing to consider—achievements from technologically less visible or glamorous branches of engineering, the project was designed to be a collaboration between the NAE and professional engineering organizations, including among others those representing civil, mechanical, chemical, and electrical engineers. In all, more than sixty engineering societies were asked to participate in the first phase of the project, which involved each society nominating up to five candidates for the greatest engineering achievements of the century. The grand list, based on the criterion of effect on quality of life, was determined by an anonymous panel of NAE members representing the various engineering disciplines and was released during Engineers Week 2000. The list, which comprised categories of achievement rather than specific milestones or monuments, is as follows:

1. Electrification
2. Automobile
3. Airplane
4. Water Supply and Distribution
5. Electronics
6. Radio and Television
7. Agricultural Mechanization
8. Computers
9. Telephone
10. Air Conditioning and Refrigeration
11. Highways
12. Spacecraft
13. Internet
14. Imaging
15. Household Appliances
16. Health Technologies
17. Petroleum and Petrochemical Technologies

18. Laser and Fiber Optics
19. Nuclear Technologies
20. High-performance Materials

The list is a striking reminder that the world of the late 1890s and early 1900s was quite different from that of 2000. Engineers' dreams had turned into engineers' realities. Chances are that our ancestors then did not yet have electricity in their homes; did not have a car; had not even heard of an airplane; drank untreated water; had no electronic devices for calculation or amusement; and had yet to hear a radio broadcast or see a television show. If they lived on a farm—as so many then did—they would not have had a tractor or other labor-saving machinery. If they had heard the word "computer," it referred to a person—usually a woman—who carried out calculations by longhand. A telephone was a luxury, and even at mid-century long-distance calls were reserved for emergencies and very special events. There was no air-conditioning or refrigeration as we know them today. Highways were muddy, rutted roads in which wagons frequently got stuck. Spacecraft were science fiction; the Internet was not even that; and imaging meant the mysterious new X-rays. What household appliances there were were powered by strong backs. Health technologies were primitive by today's standards. The petroleum industry was in its infancy; plastic convenience items were rare and expensive. Lasers and fiber optics, nuclear power, and high-performance materials like nylon and Kevlar were undreamed of.

Predicting the technological future has always been risky business, for the world of invention and engineering never ceases to push the limits of technology to come up with surprises that surprise even the experts. In 1899, it has been reported, the U.S. commissioner of patents expected his office to soon be obsolete, supposedly believing that "everything that can be invented has been invented." In 1901, Wilbur Wright confessed to his brother Orville the belief "that man would not fly for fifty years." In 1903, a bank president advised Henry Ford's lawyer not to invest in the Ford Motor Company, arguing, "The horse is here to stay, but the automobile is only a novelty—a fad." As late as 1967, Lee De Forest, considered a father of radio, asserted that man would never reach the moon, "regardless of all future scientific advances." And in 1977, the president and founder of Digital Equipment Corporation stated, "There is no reason for any individual to have a computer in their home." Fortunately, not all inventors and engineers listen to predictions, even those made by experts; they just push the limits of technology farther and farther into the future.

Regardless of what may be thought of millennium madness, anniversaries, outstanding lists, predictions, and the like, the naming of the greatest engineering achievements provided an opportunity for all the beneficiaries of these achievements to reflect on the nature and importance of engineering and its virtually endless, if not countless, contributions to making daily life so much more comfortable and convenient than it must have been a century ago, let alone a millennium ago. In many cases, twentieth-century engineers pushed the limits of technology to accomplish things that were not even dreamed of in the nineteenth century. And so it will be in the twenty-first century, with the contents of any list of engineering achievements that will be compiled in the late 2090s being virtually unpredictable today.

Acknowledgments and Bibliography

These essays appeared first in *American Scientist*, for which I have been writing the "Engineering" column since 1991. I am grateful to the magazine's editor, Rosalind Reid, and to its managing editor, David Schoonmaker, who also copyedited most of the original manuscripts for these essays, for providing me this continuing forum. As usual, I am indebted to numerous librarians, especially those at Duke University, and in particular Eric Smith and Linda Martinez, for their responsiveness to my many requests and queries. I am also grateful to the readers of *American Scientist* who have written a steady stream of warm and encouraging letters, often with addenda to, amplifications on, and corrections of some of my assertions and speculations. I am further indebted to a variety of individuals for help on specific essays. These are noted in the appropriate place below, followed by brief bibliographies of sources and further reading.

As always, I owe considerable thanks to my wife, Catherine Petroski, who continues to be my first reader and always provides invaluable early feedback. In this case, I am also grateful to her for providing some illustrations and for digitizing the others. Finally, but certainly not least of all, I am thankful for the care taken at Alfred A. Knopf in preparing the manuscript for the press. I am especially indebted to production editor Ellen Feldman and copyeditor Timothy Mennel. They, like my editor and friend Ashbel Green, provided valuable feedback on the manuscript and saved me the embarrassment of some infelicities of style and a lingering garrulousness.

Dimensions of bridges, buildings, and other structures are given in feet or meters, as appropriate to the historical, national, or technological context in which they are discussed. In some cases, when bridges from different eras and cultures are compared, all dimensions are given in the one set of units that seems to be more natural for the comparison. For those unaccustomed to one or the other of these length measures, a meter is approximately three feet (3.3 to be more exact), and conversely a foot is approximately three-tenths of a meter.

ART IN IRON AND STEEL

First appeared in *American Scientist* for July–August 2002 under the title "Art and Iron and Steel." The essay was adapted from the plenary lecture given by the author on November 8, 2001, at the Sigma Xi Forum on Science, the Arts, and the Humanities held in Raleigh, North Carolina.

Harriss, Joseph. *The Tallest Tower: Eiffel and the Belle Epoque.* Boston: Houghton Mifflin, 1975.

Shapiro, Mary J. *A Picture History of the Brooklyn Bridge.* New York: Dover Publications, 1983.

BRIDGES OF AMERICA

First appeared in *American Scientist* for May–June 1996.

Annan, Jason, and Pamela Gabriel. *The Great Cooper River Bridge.* Columbia: University of South Carolina Press, 2002.

Cortright, Robert S. *Bridging.* Tigard, Ore.: Bridge Ink, 1994.

———. *Bridging: Discovering the Beauty of Bridges.* Tigard, Ore.: Bridge Ink, 1998.

DeLony, Eric. *Landmark American Bridges.* New York: American Society of Civil Engineers, 1993.

Edwards, Llewellyn Nathaniel. *A Record of History and Evolution of Early American Bridges.* Orono, Maine: University Press, 1959.

Hauck, George F. W. *The Builders of the Bridge.* Kansas City: Kansas City Section, American Society of Civil Engineers, 1995.

Jackson, Donald C. *Great American Bridges and Dams.* Washington, D.C.: Preservation Press, 1988.

Smith, Dwight A., James B. Norman, and Peter T. Dykman. *Historic Highway Bridges of Oregon.* 2nd ed. Portland: Oregon Historical Society Press, 1989.

BENJAMIN FRANKLIN BRIDGE

First appeared in *American Scientist* for September–October 2002. This essay grew out of a shorter one prepared in conjunction with the exhibition *Suspended in Time: The Benjamin Franklin Bridge,* curated by Nancy Maguire and presented in May and June 2002 at the Center for the Arts on the Camden campus of Rutgers University.

"Bridging the Delaware River at Philadelphia—Part I." *Engineering News-Record,* September 30, 1926, pp. 530–35. "Part II," ibid., October 7, 1926, pp. 578–84.

Carswell, Charles. *The Building of the Delaware River Bridge Connecting Philadelphia, Pa., and Camden, N.J.* Burlington, N.J.: Enterprise Publishing, 1926.

"Engineering Studies for Philadelphia-Camden Bridge." *Engineering News-Record,* June 23, 1921, pp. 1086–90.

Petroski, Henry. *Engineers of Dreams: Great Bridge Builders and the Spanning of America.* New York: Alfred A. Knopf, 1995.

FLOATING BRIDGES

First appeared in *American Scientist* for July–August 2003. I am grateful to several readers who, over the years, have provided me with information and encouraged me to write about floating bridges. Among those I am especially indebted to are David J. Engel of Houston, Bruce F. Curtis of Boulder, and M. Myint Lwin formerly of the Washington State Department of Transportation and now with the Federal Highway Administration. I am also grateful to Bill Halpern for bringing to my attention the draw boats sunk in Lake Champlain and for directing me to the extant floating bridge at Brookfield, Vermont. Finally, I owe thanks to several readers of my column who wrote to me shortly after this one appeared: David Ferster and Phillip Hartley Smith first brought to my attention the floating bridges in Willemstad and Hobart, respectively, and Theodore Katsanis corrected my description of the Hood Canal bridge.

Esser, Phillip Seven. "Hadley, Homer More (1885–1967), Engineer." *The Online Encyclopedia of Seattle/King County History:* www.historylink.org/output.cFM?file_id=4004.

Kuesel, Thomas R. "Floating Bridge for 100 Year Storm." *Civil Engineering,* June 1985, pp. 61–65.

Lwin, M. Myint. "The Lacey V. Murrow Floating Bridge, USA." *Structural Engineering International* 3 (1993): 145–48.

Maruyama, Tadaaki, Eiichi Watanabe, and Hiroshi Tanaka. "Floating Swing Bridge with a 280 m Span, Osaka." *Structural Engineering International* 8 (1998): 174–75.

Watanabe, E. "Floating Bridges: Past and Present." *Structural Engineering International* 13 (2003): 128–32.

CONFEDERATION BRIDGE

This essay, under the title "The Fixed Link," first appeared in *American Scientist* for January–February 1997. I am grateful to Peter Moon of Toronto for sending me newspaper stories on the progress of Confederation Bridge throughout its construction, and to Nicole Phillips for the image of the completed bridge.

Begley, Lorraine, ed. *Crossing That Bridge: A Critical Look at the PEI Fixed Link.* Charlottetown, P.E.I.: Ragweed Press, 1993.

Campbell, Murray. "PEI Bridge Builds Controversy." Toronto *Globe and Mail,* June 5, 1996, pp. A1, A10. See also p. D8.

Cortright, Robert S. *Bridging.* Tigard, Ore.: Bridge Ink, 1994.

———. *Bridging: Discovering the Beauty of Bridges.* Tigard, Ore.: Bridge Ink, 1998.

Gilmour, Ross, et al. "Northumberland's Ice Breaker." *Civil Engineering,* January 1997, pp. 34–38.

Peer, George A. "Getting Ready to Erect the Big Ones." *Heavy Construction News,* January 1995. (See also various subsequent issues.)

Strait Crossing Development. "Northumberland Strait Bridge Project." *Nova Scotia Business Journal,* November 1995, advertising supplement.

Tadros, Gamil. "The Design and Construction of the Northumberland Strait Fixed Link Project." *Canadian Civil Engineer,* September 1995, pp. 18–21.

PONT DE NORMANDIE

First appeared in *American Scientist* for September–October 1995.

Pacheco, Benito M., and Yozo Fujino. "Keeping Cables Calm." *Civil Engineering,* October 1993, pp. 56–58.

Reina, Peter. "Up and Away to a World Record." *Engineering News-Record,* September 19, 1994, pp. 76–82.

———. "Award of Excellence: Virlogeux." *Engineering News-Record,* February 20, 1995, pp. 32–38.

Virlogeux, M. "Normandie Bridge Design and Construction." *Proceedings of the Institution of Civil Engineers, Structures and Buildings* 99 (1993): 281–302. See also the discussion in ibid., 104 (1994): 357–60.

———. "The Normandie Bridge, France: A New Record for Cable-Stayed Bridges." *Structural Engineering International* 4 (1994): 208–13.

Winney, Mike. "Dampers to Cut Normandie Vibrations." *New Civil Engineer,* December 15–22, 1994, pp. 4–5.

BRITANNIA BRIDGE

This essay, under the title "The Britannia Tubular Bridge," first appeared in *American Scientist* for May–June 1992.

Clark, Edwin. *The Britannia and Conway Tubular Bridges.* London: Day and Son, 1850.

Fairbairn, William. *An Account of the Construction of the Britannia and Conway Tubular Bridges.* London: John Weale, 1849.

Rolt, L. T. C. *George and Robert Stephenson: The Railway Revolution.* London: Longmans, 1960.

Rosenberg, N., and W. G. Vincenti. *The Britannia Bridge: The Generation and Diffusion of Technological Knowledge,* Cambridge, Mass.: MIT Press, 1978.

TOWER BRIDGE

First appeared in *American Scientist* for March–April 1995.

Forward, David C. "Designing London's Tower Bridge." *Mechanical Engineering,* February 1995, pp. 80–83.

"The Tower Bridge." *Engineering* 56 (1893): 352–55.

"The Tower Bridge, London, England." *Engineering News* 31 (1894): 43–47.

Waddell, J. A. L. *Bridge Engineering.* Volume 2. New York: John Wiley and Sons, 1916.

DRAWING BRIDGES

First appeared in *American Scientist* for July–August 1999. I am grateful to Robert J. Healy, assistant deputy chief engineer of the Office of Bridge Development of the Maryland Department of Transportation, who called my attention to the Severn River and Wilson bridge competitions and who provided invaluable information about them.

Billington, David P. *The Tower and the Bridge: The New Art of Structural Engineering.* New York: Basic Books, 1983.

Maryland Department of Transportation. *Aesthetic Bridges: Users Guide.* Baltimore: State Highway Administration, 1993.

―――. *Notice to Architects and Engineers: Request for Expressions of Interest: Woodrow Wilson Bridge Replacement.* Annapolis: State Highway Administration, 1998.

Severn River Bridge Design Competition: Program and Rules. Annapolis: Maryland State Highway Administration and Governor's Office of Art and Culture, 1989.

Woodrow Wilson Bridge Web site: www.wilsonbridge.com.

AN EYE-OPENING BRIDGE

This essay, under the title "Design Competition," first appeared in *American Scientist* for November–December 1997. I am grateful to Chris Jeffrey, who kindly provided me with a copy of the documentation for the design competition and information on its outcome, and to Sally Cowell of Wilkinson Eyre Architects for her patience and persistence in supplying me with images of the firm's bridges.

Barbey, M. F. *Civil Engineering Heritage: Northern England.* London: Thomas Telford, 1981.

Beckett, Derrick. *Stephensons' Britain.* Newton Abbot, Devon, England: David and Charles, 1984.

Billington, David P. "Aesthetics in Bridge Design— The Challenge." In *Bridge Design: Aesthetics and Developing Technologies,* ed. Adele Fleet Bacow and Kenneth

E. Kruckemeyer. Boston: Massachusetts Department of Public Works and Massachusetts Council on the Arts and Humanities, 1986. Pp. 3–16.

Brown, Jeff L. "Poole Harbour Span Design Sets Sail." *Civil Engineering,* March 2003, p. 13.

Clark, G. M., and J. Eyre. "The Gateshead Millennium Bridge." *The Structural Engineer* 79 (2001): 30–35.

Haldane, J. W. C. *Civil and Mechanical Engineering: Popularly and Socially Considered.* London: E. and F. N. Spon, 1890.

Jeffrey, Chris. "A New Bridge Across the Tyne: Competition Documentation." Gateshead, Tyne and Wear, England: Gateshead Metropolitan Borough Council, 1996.

Loukaitou-Sideris, A., and T. Banerjee. "The Anatomy of a Design Competition: An Inquiry." In *Proceedings of the 1987 Conference on Planning and Design in Architecture,* ed. J. P. Protzen. New York: American Society of Mechanical Engineers, 1987. Pp. 29–34.

Pope, Chris. "Raising Eyebrows." *Professional Engineering,* November 28, 2001, pp. 23–25.

Wilkinson, Chris, and Jim Eyre. *Bridging Art and Science.* London: Booth-Clibborn Editions, 2001.

MILLENNIUM LEGACIES

First appeared in *American Scientist* for September–October 2001.

Comptroller and Auditor General. *The Millennium Dome.* Report ordered by the House of Commons. London: The Stationery Office, 2000.

Dallard, Pat, et al. "London Millennium Bridge: Pedestrian-Induced Lateral Vibration." *Journal of Bridge Engineering* 6 (2001): 412–17.

Day, Martyn. "Marks Barfield Creates a Towering Eyeful for Central London." *MSM,* May 2000, cover story: archive.msmonline.com/old_archive/2000/05/cover2.htm.

Mann, Allan, special issue ed. "The British Airways London Eye." *The Structural Engineer* 79 (2001): 15–35.

[Millennium Bridge Trust]. *Blade of Light: The Story of London's Millennium Bridge.* London: Penguin Press, 2001.

Nakamura, Shun-ichi, and Yozo Fujino. "Lateral Vibration on a Pedestrian Cable-Stayed Bridge." *Structural Engineering International* 12 (2002): 295–300.

Rattenbury, Kester. *The Essential Eye: British Airways London Eye.* London: HarperCollins, 2002.

Reina, Peter, and Aileen Cho. "Spans Sway Underfoot in Europe." *Engineering News-Record,* July 10, 2000, pp. 14–15.

Waters, Tony. *Bridge by Bridge Through London: The Thames from Tower Bridge to Teddington.* London: Pryor Publications, 1989.

BROKEN BRIDGES

First appeared in *American Scientist* for July–August 1994.

NEW AND FUTURE BRIDGES

First appeared in *American Scientist* for November–December 1998. I am grateful to Charles Seim for providing the illustration of the proposed Gibraltar bridge.

Brown, David J. *Bridges.* New York: Macmillan, 1963.

Cole, Terrence. "The Bridge to Tomorrow: Visions of the Bering Strait Bridge." *Alaska History* 5 (1990): 1–15.

Kashima, Staoshi, and Makoto Kitagawa. 1997. "The Longest Suspension Bridge." *Scientific American,* December 1997, pp. 89–94.

Leto, I. V. "Preliminary Design of the Messina Strait Bridge." *Proceedings of the Institution of Civil Engineers* 102 (1994): 122–29. See also the discussion in ibid., 108 (1994): 39–42.

Lin, T. Y., and Philip Chow. "Gibraltar Strait Crossing—A Challenge to Bridge and Structural Engineers." *Structural Engineering International* 1 (1991): 53–58.

Marquez, Michael, et al. "New Carquinez Strait Suspension Bridge, San Francisco, California." *Structural Engineering International* 13 (2003): 100–102.

Menn, Christian, and David P. Billington. "Breaking Barriers of Scale: A Concept for Extremely Long Span Bridges." *Structural Engineering International* 5 (1995): 48–50.

Reina, Peter. "For Historic Link, Team Focuses on Earlier Experience." *Engineering News-Record,* May 1996, pp. 24–28.

DORTON ARENA

First appeared in *American Scientist* for November–December 2002, on the occasion of the fiftieth anniversary of the Dorton Arena and of its dedication as a National Historic Civil Engineering Landmark. I was aided greatly in writing this essay by reference materials accompanying the nomination of the arena as a landmark submitted to the American Society of Civil Engineers' History and Heritage Committee, which I chair. The nomination was submitted by David B. Peterson, chairman of the North Carolina Section History and Heritage Committee, and much of the supplemental reference material for the nomination was provided by the Linda Hall Library in Kansas City, Missouri. I am also grateful to Heather Overton, public information officer for the North Carolina State Fair, for providing further information.

Addis, William. "Design Revolutions in the History of Tension Structures." *Structural Engineering Review* 6 (1994): 1–10.

Bradshaw, Richard, et al. "Special Structures: Past, Present, and Future." *Journal of Structural Engineering* 128 (2002): 691–709.

Krishna, Prem. *Cable-Suspended Roofs.* New York: McGraw-Hill, 1978.

Merritt, Frederick S. "Curved Roof on Cables Spans Big Arena." *Engineering News-Record,* February 5, 1953, pp. 31–37.

Otto, Frei, and Friedrich-Karl Schleyer. *Tensile Structures.* Volume 2. Cambridge, Mass.: MIT Press, 1969.

"Parabolic Pavilion." *Architectural Forum,* October 1952, pp. 134–39.

"Permanent State Exposition Proposed." [North Carolina] *Agricultural Review,* November 1948, p. 2.

Schafer, Bruce Harold. *The Writings and Sketches of Matthew Nowicki.* Charlottesville: University Press of Virginia, 1973.

Severud, Fred N. "Materials Combined to Advantage—Concrete in Compression, Steel in Tension." *Civil Engineering,* March 1954, pp. 52–55.

Waugh, Elizabeth Culbertson. "Firm in an Ivied Tower." *North Carolina Architect,* January–February 1971, pp. 9–28.

BILBAO

First appeared in *American Scientist* for July–August 1998. I am grateful to Robert Sinn, of Skidmore, Owings & Merrill, for providing images of the structure.

Barrenecha, Raul A. "Gehry's Guggenheim." *Architecture,* September 1996, pp. 177–79.

Churchill, Bonnie. "Laying the Cornerstone for a City's Dream." *Christian Science Monitor,* July 2, 1997, p. 13.

Iyengar, Hal, Lawrence Novak, Robert Sinn, and John Zils. "Framing a Work of Art." *Civil Engineering,* March 1998, pp. 44–47.

LeCuyer, Annette. "Building Bilbao." *Architectural Review,* December 1997, pp. 4–5.

Plaza, Beatriz. "Evaluating the Influence of a Large Cultural Artifact in the Attraction of Tourism: The Guggenheim Museum Bilbao Case." *Urban Affairs Review* 36 (2000): 264–74.

Reina, Peter. "Museum's Modular Steel Grid Sets Free-Form Surfaces Straight." *Engineering News-Record,* October 13, 1997, pp. 60–62.

Riding, Alan. "The Basques Get Modern: A Gleaming New Guggenheim for Grimy Bilbao." *New York Times,* June 24, 1997, p. C9.

Tomkins, Calvin. "The Maverick." *The New Yorker,* July 7, 1997, pp. 38–45.

SANTIAGO CALATRAVA

First appeared in *American Scientist* for March–April 1997. I am grateful to Diana Redlich of Graef, Anhalt, Schloemer & Associates for providing images of the addition to the Milwaukee Art Museum.

Forgey, Benjamin. "Winged Victory." *Washington Post,* November 4, 2001, p. G1.

Fowler, David. "Artisantiago Calatrava." *New Civil Engineer,* October 8, 1992, pp. 20–21.

Frampton, Kenneth, and Anthony C. Webster. *Santiago Calatrava: Bridges.* Zurich: Artemis, 1993.

Harbison, Robert. *Creatures from the Mind of the Engineer: The Architecture of Santiago Calatrava.* Zurich: Verlag für Architektur, 1992.

Kamin, Blair. "Santiago Calatrava Marries Sculpture and Structure." *Architectural Record,* March 2002, p. 92.

Metz, Tracy. "Structural Dynamics." *Architectural Record,* October 1990, pp. 54–61.

———. "Express Track." *Architectural Record,* August 1991, pp. 84–89.

Pollalis, Spiro N. *What Is a Bridge? The Making of Calatrava's Bridge in Seville.* Cambridge, Mass.: MIT Press, 2002.

Schumaker, Mary Louise. "Solution Made for Shade." *Milwaukee Journal Sentinel,* May 22, 2001, p. 1A.

Stewart, Doug. "Transforming the Beauty of Skeletons into Architecture." *Smithsonian,* November 1996, pp. 76–86.

Tonkin, Boyd. "Travelling Hopefully." *New Statesman and Society,* March 18, 1994, pp. 49–50.

Webster, Anthony C. "Utility, Technology, and Expression." *Architectural Review,* November 1992, pp. 68–74.

FAZLUR KHAN

First appeared in *American Scientist* for January–February 1999. I am grateful to Robert B. Johnson, longtime public-relations chairman for the Structural Engineers Association of Illinois, for providing me over the years with information on the Fazlur Khan sculpture and street naming. I am also grateful to John Zils, associate partner at Skidmore, Owings & Merrill, for further information on the sculpture's location in the Sears Tower.

Ali, Mir M. *Art of the Skyscraper: The Genius of Fazlur Khan.* New York: Rizzoli, 2001.

[Bangladesh Association of Greater Chicagoland]. *"Fazlur R. Khan Way" Dedication Ceremony.* Program. June 16, 1998.

"Construction's Man of the Year: Fazlur R. Khan." *Engineering News-Record,* February 10, 1972, pp. 20–21, 23, 25.

Gapp, Paul. "A Chicago Giant Finds His Place." *Chicago Tribune,* April 20, 1989, sec. 5, p. 9.

Khan, Fazlur R. "The John Hancock Center." *Civil Engineering,* October 1967, pp. 38–42.

[Morrison, Allen]. "Top Foreign-Born Civil Engineers Speak Their Minds." *Civil Engineering,* October 1980, pp. 114–20.

Mufti, Aftab A., and Baidar Bakht. "Fazlur Khan (1929–1982): Reflections on His Life and Works." *Canadian Journal of Civil Engineering* 29 (2002): 238–45.

Thornton, Charles H., et al. *Exposed Structure in Building Design.* New York: McGraw-Hill, 1993.

THE FALL OF SKYSCRAPERS

First appeared in *American Scientist* for January–February 2002. Some of the ideas in the essay were first published in the Outlook section of the *Washington Post* for September 16, 2001.

Bascomb, Neal. *Higher: A Historic Race to the Sky and the Making of a City.* New York: Doubleday, 2003.

Bazant, Zdenek P., and Yong Zhou. "Why Did the World Trade Center Collapse?—Simple Analysis." *Journal of Engineering Mechanics* 128 (2002): 2–6. See also www3.tam.uiuc.edu/news/200109wtc/.

Darton, Eric. *Divided We Stand: A Biography of New York's World Trade Center.* New York: Basic Books, 1999.

Gillespie, Angus Kress. *Twin Towers: The Life of New York City's World Trade Center.* New York: New American Library, 2002.

Mackin, Thomas J. "Engineering Analysis of Tragedy at WTC." Presentation slides for ME 346, Department of Mechanical Engineering, University of Illinois at Urbana-Champaign, 2001.

VANITIES OF THE BONFIRE

First appeared in *American Scientist* for November–December 2000. This essay also includes material from a shorter one that appeared in the Outlook section of the *Washington Post* for November 28, 1999. I am grateful to Dave Amber of *The Battalion,* the Texas A&M student newspaper, for alerting me to the release of the special commission's final report and for providing me with a copy of it. William W. Ward, P.E., a 1948 graduate of the Agricultural and Mechanical College of Texas, was kind enough to read portions of this essay in manuscript. I am grateful to Steven Smith, director of the Cushing Memorial Library and Archives at Texas A&M, for bringing to my attention its extensive collection of Bonfire photos, and to Lacey Vaculin for scanning some of them for me.

Special Commission on the 1999 Texas A&M Bonfire. *Final Report.* [College Station: Texas A & M University], 2000. See www.tamu.edu/bonfire-commission/reports.

Tang, Irwin A. *The Texas Aggie Bonfire: Tradition and Tragedy at Texas A&M.* [Austin, Tex.: The It Works, 2000.]

ST. FRANCIS DAM

First appeared in *American Scientist* for March–April 2003. I am grateful to Jon Wilkman for sharing with me his proposal to make a documentary film on the St.

Francis Dam disaster, for providing background material, and for encouraging me to write on the subject. (He is also my source for the Mulholland quote from the coroner's inquest.) Among the invaluable resources provided by Wilkman was a CD-ROM of J. David Rogers's research on the failure of the dam. Rogers's paper in the book edited by Doyce Nunis, referenced below, was another valuable source of information and quotations on the history of the project. I am also indebted to Irving Sherman, for his insight into the nature of materials and the density of sediment. Finally, I am grateful to D. C. Jackson for providing background material on the state of the art in dam design and construction in the early twentieth century, and to Norris Hundley for including me in e-mail exchanges regarding the history of the dam, its failure, and the aftermath.

American Society of Civil Engineers, Committee of Board of Direction. "Essential Facts Concerning the Failure of the St. Francis Dam." *Proceedings of the ASCE* 55 (1929): 2147–63.

Bowers, Nathan A. "St. Francis Dam Catastrophe—A Great Foundation Failure." *Engineering News-Record*, March 22, 1928, pp. 466–72.

———. "St. Francis Dam Catastrophe—A Review Six Weeks After." *Engineering News-Record*, May 10, 1928, pp. 726–33.

Clements, Thomas. "St. Francis Dam Failure of 1928." In *Engineering Geology in Southern California*, ed. Richard Lung and Richard Proctor. Los Angeles: Association of Engineering Geologists, 1966. Pp. 88–91.

Grunsky, C. E., and E. L. Grunsky, "St. Francis Dam Failure." *Western Construction News*, May 25, 1928, pp. 314–24.

Jackson, D. C. *Building the Ultimate Dam: John S. Eastwood and the Control of Water in the West.* Lawrence: University Press of Kansas, 1995.

Lawrance, Charles H. *The Death of the Dam: A Chapter in Southern California History.* Privately printed, 1995.

Nunis, Doyce B., Jr., ed. *The St. Francis Dam Disaster, Revisited.* Los Angeles and Ventura: Historical Society of Southern California and Ventura County Museum of History and Art, 1995.

Outland, Charles F. *Man-Made Disaster: The Story of St. Francis Dam, Its Place in Southern California's Water System, Its Failure, and the Tragedy of March 12 and 13, 1928, in the Santa Clara River Valley.* Glendale, Calif.: Arthur H. Clark, 1963.

Rogers, J. David, "A Man, a Dam, and a Disaster: Mullholland and the St. Francis Dam." In Nunis, pp. 1–109.

———. "Man Made Disaster at an Old Landslide Dam Site." Privately distributed CD-ROM, 2000.

THREE GORGES DAM

First appeared in two parts, under the titles "China Journal I" and "China Journal II," in *American Scientist* for May–June and July–August 2001. My visit to China was made possible by the invitation of the People to People Ambassadors Program to lead a civil-engineering delegation focused on the Three Gorges dam project.

The official program of the delegation began in Los Angeles on November 14, 2000, and ended in Beijing on November 27.

Barber, Margaret, and Grainne Ryder, eds. *Damming the Three Gorges: What Dam Builders Don't Want You to Know.* 2nd ed. London: Earthscan Publications, 1993.

Dai Qing. *The River Dragon Has Come! The Three Gorges Dam and the Fate of China's Yangtze River and Its People.* Armonk, N.Y.: M. E. Sharpe, 1998.

Hersey, John. *A Single Pebble.* New York: Alfred A. Knopf, 1956.

Zhu Rulan et al. "The Three Gorges Project: Key to the Development of the Yangtze River." *Civil Engineering Practice,* Spring/Summer 1997, pp. 39–72.

FUEL CELLS

First appeared in *American Scientist* for September–October 2003. I am grateful to Daniel Muzyka, dean of commerce at the University of British Columbia, who invited and encouraged me to join the Chrysalix Industry Advisory Committee, which he chaired. I am grateful to Wal van Lierop, president and CEO of Chrysalix, and to board chairman Michael J. Brown, both of whom read an early manuscript of this column and provided helpful feedback on the history, status, and future of fuel cells.

Hoffmann, Peter. *Tomorrow's Energy: Hydrogen, Fuel Cells, and the Prospects for a Cleaner Planet.* Cambridge, Mass.: MIT Press, 2001.

Koppel, Tom. *Powering the Future: The Ballard Fuel Cell and the Race to Change the World.* Toronto: John Wiley and Sons Canada, 1999.

Westbrook, Michael H. *The Electric Car: Development and Future of Battery, Hybrid and Fuel-Cell Cars.* London: Institution of Electrical Engineers, 2001.

ENGINEERS' DREAMS

First appeared in *American Scientist* for July–August 1997.

Asimov, Isaac. *Asimov's Biographical Encyclopedia of Science and Technology.* 2nd ed. Garden City, N.Y.: Doubleday, 1982.

Contemporary Authors. Vols. 9–12, first revision, and vol. 113. Detroit: Gale Research, 1974, 1985.

Current Biography. New York: H. W. Wilson, 1953, 1957.

Davidson, Frank P. *Macro: A Clear Vision of How Science and Technology Will Shape Our Future.* New York: Morrow, 1983.

Kuehl, Warren F. *Biographical Dictionary of Internationalists.* Westport, Conn.: Greenwood Press, 1983.

Ley, Willy. *Engineers' Dreams.* New York: Viking Press, 1954; rev. ed., 1964.

Taylor, Michael. "The Dreamers." *San Francisco Chronicle,* September 8, 1996.

ENGINEERS' ACHIEVEMENTS

This essay, under the title "Time-Sensitive Material," first appeared in *American Scientist* for January–February 2000. As originally published, the essay did not include the final list of great engineering achievements of the twentieth century chosen by the National Academy of Engineering committee of which I was a member. (Committee membership was made public with the announcement of the list.) I am grateful to Edward Allen for sending me the *Newsweek* page of wrongheaded predictions.

"Cloudy Days in Tomorrowland." *Newsweek,* July 27, 1997, p. 86.

Constable, George, and Bob Somerville. *A Century of Innovation: Twenty Engineering Achievements That Transformed Our Lives.* Washington, D.C.: Joseph Henry Press, 2003.

Gould, Stephen Jay. *Questioning the Millennium: A Rationalist's Guide to a Precisely Arbitrary Countdown.* New York: Harmony Books, 1999.

National Academy of Engineering. *Ten Outstanding Achievements, 1964–1989: Engineering and the Advancement of Human Welfare.* Washington, D.C.: NAE, 1989.

———. *Greatest Engineering Achievements of the 20th Century.* 2000. www.greatachievements.org.

"125 Years of Top Projects." *Engineering News-Record,* July 26, 1999, p. 41.

Society for the Promotion of Engineering Education. *Proceedings of the Thirty-eighth Annual Meeting,* Montreal, Quebec, June 26–28, 1930.

"Top People of the Past 125 Years." *Engineering News-Record,* August 30, 1999, pp. 27–54.

List of Illustrations and Credits

Index

Page numbers in *italics* refer to illustrations.

273

A NOTE ON THE TYPE

This book was set in Minion, a typeface produced by the Adobe Corporation specifically for the Macintosh personal computer and released in 1990. Designed by Robert Slimbach, Minion combines the classic characteristics of old-style faces with the full complement of weights required for modern typesetting.

Composed by North Market Street Graphics, Lancaster, Pennsylvania
Printed and bound by Berryville Graphics, Berryville, Virginia
Designed by Robert C. Olsson